THIRD EDITION

CASE FILES®:
Anatomy

Eugene C. Toy, MD
Vice Chair of Academic Affairs and
  Program Director
Houston Methodist Hospital
Obstetrics and Gynecology Residency
  Program
Houston, Texas
Clinical Professor and Clerkship Director
Department of Obstetrics and Gynecology
University of Texas Medical School at
  Houston
Houston, Texas
John S. Dunn Senior Academic Chair
St. Joseph Medical Center
Houston, Texas

Lawrence M. Ross, MD, PhD
Adjunct Professor
Department of Neurobiology and
  Anatomy
University of Texas Medical School at
  Houston
Houston, Texas

Han Zhang, MD
Associate Professor, Research
Department of Neurobiology and
  Anatomy
University of Texas Medical School at
  Houston
Houston, Texas

Cristo Papasakelariou, MD, FACOG
Clinical Professor,
Department of Obstetrics and Gynecology
University of Texas Medical Branch
Galveston, Texas
Clinical Director of Gynecologic Surgery
St. Joseph Medical Center
Houston, Texas

McGraw Hill Education | Medical

New York   Chicago   San Francisco   Athens   London   Madri
Mexico City   Milan   New Delhi   Singapore   Sydney   Toronto

Case Files®: Anatomy, Third Edition

1 2 3 4 5 6 7 8 9 0   DOC/DOC   19 18 17 16 15 14

ISBN 978-0-07-179486-2
MHID 0-07-179486-7

Notice

Medicine is an ever-changing science. As new research and clinical experience broaden our knowledge, changes in treatment and drug therapy are required. The authors and the publisher of this work have checked with sources believed to be reliable in their efforts to provide information that is complete and generally in accord with the standard accepted at the time of publication. However, in view of the possibility of human error or changes in medical sciences, neither the editors nor the publisher nor any other party who has been involved in the preparation or publication of this work warrants that the information contained herein is in every respect accurate or complete, and they disclaim all responsibility for any errors or omissions or for the results obtained from use of the information contained in this work. Readers are encouraged to confirm the information contained herein with other sources. For example and in particular, readers are advised to check the product information sheet included in the package of each drug they plan to administer to be certain that the information contained in this work is accurate and that changes have not been made in the recommended dose or in the contraindications for administration. This recommendation is of particular importance in connection with new or infrequently used drugs.

This book was set in Goudy by Cenveo® Publisher Services.
The editors were Catherine A. Johnson and Cindy Yoo.
The production supervisor was Catherine Saggese.
Project management was provided by Anupriya Tyagi, Cenveo Publisher Services.
The cover designer was Thomas De Pierro.
RR Donnelley was printer and binder.

This book is printed on acid-free paper.

Toy, Eugene C., author.
    Case files. Anatomy / Eugene C. Toy, Lawrence M. Ross, Han Zhang, Cristo Papasakelariou.—Third edition.
        p. ; cm.
    Anatomy
    Preceded by Case files. Anatomy / Eugene C. Toy ... [et al.]. 2nd ed. c2008.
    Includes bibliographical references and index.
    ISBN 978-0-07-179486-2 (alk. paper)—ISBN 0-07-179486-7 (alk. paper)
    I. Ross, Lawrence M., author.   II. Zhang, Han, 1959- author.   III. Papasakelariou, Cristo, author.   IV. Title.
V. Title: Anatomy.
    [DNLM: 1. Anatomy—Case Reports.   2. Anatomy—Problems and Exercises.   3. Clinical Medicine—
Case Reports.   4. Clinical Medicine—Problems and Exercises.   QS 18.2]
    RB25
    611—dc23   2014006471

Hitoshi "Toshi" Nikaidoh (1968–2003)

We dedicate this book to our dear friend, Dr. Toshi Nikaidoh, who led by example, always beyond the call of duty, and along the way, taught so many of us about so many important things about life.

As a surgeon-to-be, he tutored fellow lower-level medical students on not only how to master the challenges of gross anatomy but also how to develop the skillful art of dissection and respect for the human body.

As a spiritual leader, he taught his youth group not only the meaning of good fellowship by recalling good times spent on missionary travels abroad, but also the value of good worship by sharing his faith along the way.

As a physician, he taught patients not only to hope when all hope is lost but also to have faith through which peace can be found.

And as a friend, son, brother, or just that smiling doctor in the hallway with the bow tie, he taught us how truly possible it is for one person to make a world of difference.

Toshi's dedication to academics and education, his compassion for the sick and less fortunate, and his tireless devotion to his faith, family, and friends have all continued to touch and change lives of all who knew him, and even of all who only knew of him.

Miki Takase, MD
Fellow classmate
University of Texas Medical School at Houston
St. Joseph Medical Center Ob/Gyn Resident
Written on behalf of Toshi's many friends,
classmates, fellow residents, staff, and faculty at
University of Texas Medical School at Houston and
St. Joseph Medical Center

In the memory of Dr. Hitoshi Nikaidoh, who demonstrated unselfishness, love for his fellow man, and compassion for everyone around him. He is the best example of the physician healer, and we were blessed to have known him.

—ECT

To my wife, Irene; the children, Chip, Jennifer, Jocelyn, Tricia, and Trey; and the medical students, each of whom has taught me something of value.

—LMR

To my students and colleagues, who bring joy and advancement to the teaching of anatomy; and to my family for their endless support.

—HZ

To my parents Kiriaki and Alexander, and my wife Beth, for their support, love, and encouragement.

—CP

CONTENTS

Ashley L. Gunter, MD
Resident Physician in Internal Medicine
University of Texas Medical School at Houston
Houston, Texas
Rotator Cuff Injury

Konrad P. Harms, MD
Associate Program Director
Obstetrics and Gynecology Residency Program
The Methodist Hospital-Houston
Houston, Texas
Clinical Assistant Professor
Weill Cornell School of Medicine
New York, New York
Greater Vestibular Gland Abscess

Krishna B. Shah, MD
Resident in Anesthesiology
Baylor College of Medicine
Houston, Texas
Knee Injury

Shen Song
Medical Resident
Emory University School of Medicine
Atlanta, Georgia
Hydrocephalus
Knee Injury
Rotator Cuff Injury

Allison L. Toy
Senior Nursing Student
Scott & White Nursing School
University of Mary Hardin-Baylor
Belton, Texas
Primary Manuscript Reviewer

We appreciate all the kind remarks and suggestions from the many medical students over the past 5 years. Your positive reception has been an incredible encouragement, especially in light of the short life of the *Case Files®* series. In this third edition of *Case Files®: Anatomy*, the basic format of the book has been retained. Improvements were made in updating many of the chapters. New cases include hydrocephalus, knee injury, peritoneal irritation, rotator cuff injury, and thoracic outlet syndrome. We reviewed the clinical scenarios with the intent of improving them; however, their "real-life" presentations patterned after actual clinical experience were accurate and instructive. The multiple-choice questions have been carefully reviewed and rewritten to ensure that they comply with the National Board and United States Medical Licensing Examination (USMLE) format. Through this third edition, we hope that the reader will continue to enjoy learning diagnosis and management through the simulated clinical cases. It certainly is a privilege to be teachers for so many students, and it is with humility that we present this edition.

The Authors

ACKNOWLEDGMENTS

The inspiration for this basic science series occurred at an educational retreat led by Dr. Maximillian Buja, who at the time was the dean of the medical school. Dr. Buja served as Dean of the University of Texas Medical School at Houston from 1995 to 2003 before being appointed Executive Vice President for Academic Affairs. It has been such a joy to work together with Dr. Lawrence Ross, who is a brilliant anatomist and teacher, and my new scientist author Dr. Han Zhang. Sitting side by side during the writing process as they precisely described the anatomical structures was academically fulfilling, but more so, made me a better surgeon. It has been a privilege to work with Dr. Cristo Papasakelariou, a dear friend, scientist, leader, and the finest gynecological laparoscopic surgeon I know. I would like to thank McGraw-Hill for believing in the concept of teaching by clinical cases. I owe a great debt to Catherine Johnson, who has been a fantastically encouraging and enthusiastic editor. It has been amazing to work together with my daughter Allison, who is a senior nursing student at the Scott and White School of Nursing; she is an astute manuscript reviewer and already early in her career she has a good clinical acumen and a clear writing style. Dr. Ross would like to acknowledge the figure drawings from the University of Texas Medical School at Houston originally published in Philo et al., *Guide to Human Anatomy*, Philadelphia: Saunders, 1985. At Methodist Hospital, I appreciate and am grateful to Drs. Mark Boom, Alan Kaplan, and Judy Paukert. At St. Joseph Medical Center, I would like to recognize our outstanding administrators: Pat Mathews and Paula Efird. I appreciate Linda Bergstrom's advice and assistance. Without the help from my colleagues, Drs. Konrad Harms, Priti Schachel, Gizelle Brooks-Carter, John McBride, and Russell Edwards, this manuscript could not have been written. Most importantly, I am humbled by the love, affection, and encouragement from my lovely wife, Terri, and our children, Andy and his wife Anna, Michael, Allison, and Christina.

Eugene C. Toy

Mastering the diverse knowledge within a field such as anatomy is a formidable task. It is even more difficult to draw on that knowledge, relate it to a clinical setting, and apply it to the context of the individual patient. To gain these skills, the student learns best with good anatomical models or a well-dissected cadaver, at the laboratory bench, guided and instructed by experienced teachers, and inspired toward self-directed, diligent reading. Clearly, there is no replacement for education at the bench. Even with accurate knowledge of the basic science, the application of that knowledge is not always easy. Thus, this collection of patient cases is designed to simulate the clinical approach and stress the clinical relevance to the anatomical sciences.

Most importantly, the explanations for the cases emphasize the mechanisms and structure–function principles rather than merely rote questions and answers. This book is organized for versatility to allow the student "in a rush" to go quickly through the scenarios and check the corresponding answers or to consider the thought-provoking explanations. The answers are arranged from simple to complex: the bare answers, a clinical correlation of the case, an approach to the pertinent topic including objectives and definitions, a comprehension test at the end, anatomical pearls for emphasis, and a list of references for further reading. The clinical vignettes are listed by region to allow for a more synthetic approach to the material. A listing of cases is included in Section III to aid any students who desire to test their knowledge of a certain area or to review a topic including basic definitions. We intentionally used open-ended questions in the case scenarios to encourage the student to think through relations and mechanisms.

Applying Basic Sciences to Clinical Situations

Part 1. Approach to Learning

Learning anatomy consists not only in memorization but also in visualization of the relations between the various structures of the body and understanding their corresponding functions. Rote memorization will quickly lead to forgetfulness and boredom. Instead, the student should approach an anatomical structure by trying to correlate its purpose with its design. Structures that are close together should be related not only spatially but also functionally. The student should also try to project clinical significance to the anatomical findings. For example, if two nerves travel close together down the arm, one could speculate that a tumor, laceration, or ischemic injury might affect both nerves; the next step would be to describe the deficits expected on physical examination.

The student must approach the subject in a systematic manner, by studying the **skeletal** relations of a certain region of the body, the **joints**, the **muscular system**, the **cardiovascular system** (including arterial perfusion and venous drainage), the **nervous system** (such as sensory and motor neural innervations), and the **skin**. Each bone or muscle is unique and has advantages due to its structure and limitions or perhaps vulnerability to specific injuries. The student is encouraged to read through the description of the anatomical relation in a certain region, correlate illustrations of the same structures, and then try to envision the anatomy in three dimensions. For instance, if the anatomical drawings are in the coronal plane, the student may want to draw the same region in the sagittal or cross-sectional plane as an exercise to visualize the anatomy more clearly.

Part 2. Basic Terminology

Anatomical position: The basis of all descriptions in the anatomical sciences, with the head, eyes, and toes pointing forward; the upper limbs by the side with the palms facing forward; and the lower limbs together.

Anatomical planes: A section through the body, one of four commonly described planes. The **median plane** is a single vertically oriented plane dividing the body into right and left halves, whereas the **sagittal planes** are oriented parallel to the median plane but not necessarily in the midline. **Coronal planes** are perpendicular to the median plane and divide the body into anterior (front) and posterior (back) portions. **Transverse, axial, or cross-sectional planes** pass through the body perpendicular to the median and coronal planes and divide the body into upper and lower parts.

Directionality: Superior (cranial) is toward the head, whereas **inferior (caudal)** is toward the feet; **medial** is toward the midline, whereas **lateral** is away from the midline. **Proximal** is toward the trunk or attachment, whereas **distal** is away from the trunk or attachment. **Superficial** is near the surface, whereas **deep** is away from the surface.

Motion: Adduction is movement toward the midline, whereas **abduction** is movement away from the midline. **Extension** is straightening a part of the body, whereas **flexion** is bending the structure. **Pronation** is the action of rotating the palmar side of the forearm facing posteriorly, whereas **supination** is the action of rotating the palmar side of the forearm anteriorly.

Part 3. Approach to Reading

The student should **read with a purpose** and not merely to memorize facts. Reading with the goal of comprehending the relation between structure and function is one of the keys to understanding anatomy. Also, the ability to relate the anatomical sciences to the clinical picture is critical. The following seven key questions are helpful in ensuring the effective application of basic science information to the clinical setting.

1. Given the importance of a certain required function, which anatomical structure provides the ability to perform that function?

2. Given the anatomical description of a body part, what is its function?

3. Given a patient's symptoms, what structure is affected?

4. Which lymph nodes are most likely to be affected by cancer at a particular location?

5. If an injury occurs to one part of the body, what is the expected clinical manifestation?

6. Given an anomaly such as weakness or numbness, what other symptoms or signs would the patient most likely have?

7. What is the male or female homologue to the organ in question?

Let us consider these seven issues in further detail.

1. **Given the importance of a certain required function, which anatomical structure provides the ability to perform that function?**

    The student should be able to relate the anatomical structure to a function. When approaching the upper extremity, for instance, the student may begin with the statement, "The upper extremity must be able to move in many different directions to be able to reach up (flexion), reach backward (extension), reach to the side (abduction), bring the arm back (adduction), or turn a screwdriver (pronation/supination)." Because the upper extremity must move in all these directions, the joint between the trunk and arm must be very versatile. Thus, the shoulder joint is a ball-and-socket joint to allow movement in the different directions required. Further, the shallower the socket is, the more mobility the joint has. However, the versatility of the joint makes its dislocation easier.

2. **Given the anatomical description of a body part, what is its function?**

    This is the counterpart to the previous question regarding the relation between function and structure. The student should try to be imaginative and not merely accept the textbook (rote) information. One should be inquisitive, perceptive, and discriminating. For example, a student might speculate as to why bones contain marrow and are not completely solid and might theorize as follows: "The main purpose of bones is to support the body and protect various organs. If the bones were solid, they might be slightly stronger, but they would be much heavier and be a detriment to the body. Also, production of blood cells is a critical function of the body. Thus, by having the marrow within the center of the bone, the process is protected."

3. **Given a patient's symptoms, what structure is affected?**

This is one of the most critical questions of clinical anatomy. It is also one of the major questions that a clinician must answer when evaluating a patient. In clinical problem solving, the physician elicits information by asking questions (taking the history) and performing a physical examination while making observations. The history is the single most important tool for making a diagnosis. A thorough understanding of the anatomy aids the clinician tremendously because most diseases affect body parts under the skin and require "seeing under the surface." For example, a clinical observation might be: "a 45-year-old woman complains of numbness of the perineal area and has difficulty voiding." The student might speculate as follows: "The sensory innervation of the perineal area is through sacral nerves S2 through S4, and control of the bladder is through the parasympathetic nerves, also S2 through S4. Therefore, two possibilities are a spinal cord problem involving those nerve roots or a peripheral nerve lesion. The internal pudendal nerve innervates the perineal region and is involved with micturition." Further information is supplied: "The patient states that she has experienced back pain since a fall 2 weeks ago." Now the lesion can be isolated to the spine, most likely the **cauda equina** ("horsetail"), which is a bundle of spinal nerve roots traversing through the cerebrospinal fluid.

4. **Which lymph nodes are most likely to be affected by cancer at a particular location?**

The lymphatic drainage of a particular region of the body is important because cancer may spread through the lymphatics, and lymph node enlargement may result from infection. The clinician must be aware of these pathways to know where to look for metastasis (spread) of cancer. For example, if a cancer is located on the vulva labia majora (or the scrotum in the male), the most likely lymph node involved is a superficial inguinal node. The clinician would then be alert to palpating the inguinal region for lymph node enlargement, which would indicate an advanced stage of cancer and a worse prognosis.

5. **If an injury occurs to one part of the body, what is the expected clinical manifestation?**

If a laceration, tumor, trauma, or bullet causes injury to a specific area of the body, it is important to know which crucial bones, muscles, joints, vessels, and nerves might be involved. Also, an experienced clinician is aware of particular vulnerabilities. For example, the thinnest part of the skull is located in the temporal region, and underneath this is the middle meningeal artery. Thus, a blow to the temple may be disastrous. A laceration to the middle meningeal artery would lead to an epidural hematoma because this artery is located superficial to the dura and can cause cerebral damage.

6. **Given an anomaly such as weakness or numbness, what other symptoms or signs would the patient most likely have?**

   This requires a three-step process in analysis. The student must be able to (a) deduce the initial injury on the basis of clinical findings, (b) determine the probable site of injury, and (c) make an educated guess as to which other structures are in close proximity and, if injured, what the clinical manifestations would be. To develop skill in discerning these relationships, one can begin from a clinical finding, propose an anatomical deficit, propose a mechanism or location of the injury, identify another nerve or vessel or muscle in that location, propose the new clinical finding, and so on.

7. **What is the male or female homologue to the organ in question?**

   Knowledge of male–female homologous correlates is important in understanding the embryologic relations and, hence, the resultant anatomical relations because fewer structures need to be memorized, as homologous relations are easier to discern than are two separate structures. For example, the vascular supplies of homologous structures are usually similar. The ovarian arteries arise from the abdominal aorta below the renal arteries; likewise, the testicular arteries arise from the abdominal aorta.

KEY POINTS

- The student should approach an anatomical structure by visualizing the structure and understanding its function.

- A standard anatomical position is used as a reference for anatomical planes and terminology of movement.

- There are seven key questions to consider in ensuring the effective application of basic science information to the clinical arena.

REFERENCE

Moore KL, Agur AMR, Dalley AF. *Clinically Oriented Anatomy*, 6th ed. Baltimore, MD: Lippincott Williams & Wilkins, 2010.

Clinical Cases

CASE 1

A 32-year-old woman delivered a large (4800-g) baby vaginally after a somewhat difficult labor. Her prenatal course was complicated by diabetes, which developed during pregnancy. At delivery, the infant's head emerged, but the shoulders were stuck behind the maternal symphysis pubis, requiring the obstetrician to execute maneuvers to release the infant's shoulders and complete the delivery. The infant was noted to have a good cry and pink color but was not moving its right arm.

▶ What is the most likely diagnosis?
▶ What is the most likely etiology for this condition?
▶ What is the likely anatomical mechanism for this disorder?

ANSWERS TO CASE 1:

Brachial Plexus Injury

Summary: A large (4800-g) infant of a diabetic mother is delivered after some difficulty and cannot move its right arm. There is shoulder dystocia (the infant's shoulders are stuck after delivery of the head).

- **Most likely diagnosis:** Brachial plexus injury, probably Erb palsy (Duchenne-Erb paralysis)

- **Most likely etiology for this condition:** Stretching of the upper brachial plexus during delivery

- **Likely anatomical mechanism for this disorder:** Stretching of nerve roots C5 and C6 by an abnormal increase in the angle between the neck and the shoulder

CLINICAL CORRELATION

During delivery, particularly of a large infant, shoulder dystocia may occur. In this situation, the fetal head emerges, but the shoulders become wedged behind the maternal symphysis pubis. An obstetrician will use maneuvers such as flexion of the maternal hips against the maternal abdomen (McRobert maneuver) or fetal maneuvers such as pushing the fetal shoulders into an oblique position. These actions are designed to allow delivery of the fetal shoulders without excessive traction on the fetal neck. Despite such carefully executed maneuvers, infants may be born with stretch injuries to the brachial plexus, resulting in nerve palsies. The most common of these is an upper brachial plexus stretch injury, in which nerve roots C5 and C6 are affected, resulting in weakness of the infant's arm. Such injuries usually resolve spontaneously.

APPROACH TO:

The Brachial Plexus

OBJECTIVES

1. Be able to describe the spinal cord segments, named terminal branches, and motor and sensory deficits of an **upper brachial plexus injury**

2. Be able to describe the mechanism, spinal cord segments, named terminal branches, and motor and sensory deficits of a **lower brachial plexus injury**

3. Be able to describe the mechanism, spinal cord segments, named terminal branches, and motor and sensory deficits with **cord injury** of the brachial plexus

DEFINITIONS

BRACHIAL PLEXUS: A major peripheral nerve network formed by the anterior primary rami of the fifth cervical to the first thoracic spinal nerves

UPPER BRACHIAL PLEXUS INJURY: Typically involves nerve roots C5 and C6, resulting in the upper limb hanging at the side, with medial rotation and the palm facing posteriorly

LOWER BRACHIAL PLEXUS INJURY: Less common injury involving C8 through T1 and the ulnar nerve, leading to interosseous muscle atrophy and claw hand

SHOULDER DYSTOCIA: Condition whereby the fetal head delivers vaginally but the shoulders are impacted behind the maternal bony pelvis

DISCUSSION

The **brachial plexus** arises from the inferior portion of the cervical spinal cord enlargement. It is formed by the ventral **primary rami of spinal nerves C5 through C8** and most of **T1**. The network of nerves that form the brachial plexus is divided anatomically from proximal (medial) to distal (lateral) into **roots, trunks, divisions, cords,** and **terminal branches** (mnemonic: "Randy Travis drinks cold Texas beer"). The roots of the plexus emerge from between the anterior and middle scalene muscles together with the subclavian artery. Arising from the roots are branches to the **longus colli** and **scalene muscles** and the **dorsal scapular** and **long thoracic nerves.** The roots unite to form **superior, middle,** and **inferior trunks.** The **suprascapular nerve** and the nerve to the **subclavius muscle** arise from the **superior trunk.** Each trunk is divided into **anterior** and **posterior divisions,** which will innervate musculature of the anterior and posterior compartments, respectively (Figure 1-1).

The anterior divisions of the superior and middle trunks unite to form the **lateral cord,** which branches off to the **lateral pectoral nerve.** The anterior division of the inferior trunk continues distally as the medial cord, whose branches are the **medial pectoral, medial brachial cutaneous,** and **medial antebrachial cutaneous nerves.** The posterior divisions of all three trunks unite to form the **posterior cord,** and its branches are the **upper** and **lower subscapular** and **thoracodorsal nerves.** The three cords are named according to their relation to the **axillary artery,** which passes through the plexus at this level. The terminal branches of the brachial plexus are the **axillary, musculocutaneous, median, ulnar,** and **radial nerves.**

The **axillary nerve (C5** and **C6)** arises from the **posterior cord** and courses posteriorly around the **surgical neck of the humerus,** where it is at risk for injury. The **posterior circumflex humeral artery** accompanies the nerve in this course. The axillary nerve supplies the **deltoid** and **teres minor muscles,** is sensory to the skin over the lower portion of the deltoid, and is optimally tested on the "shoulder patch" portion of the upper arm. **Axillary nerve injury,** such as that due to fracture at the **surgical neck of the humerus,** results in an **inability to abduct the arm at the shoulder to a horizontal position** and in **sensory loss in the shoulder patch area** (Figure 1-2).

The **musculocutaneous nerve (C5–C7)** is the continuation of the lateral cord. It courses distally through the coracobrachialis muscle to innervate it in addition to

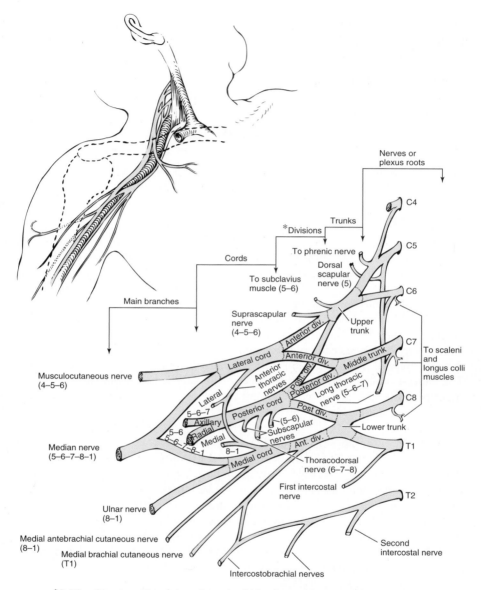

*Splitting of the plexus into anterior and posterior divisions is one of the most significant features in the redistribution of nerve fibers, because it is here that fibers supplying the flexor and extensor groups of muscles of the upper extremity are separated. Similar splitting is noted in the lumbar and sacral plexuses for the supply of muscles of the lower extremity.

Figure 1-1. The brachial plexus. (*Reproduced, with permission, from Waxman SG. Clinical Neuroanatomy, 25th ed. New York: McGraw-Hill, 2003:348.*)

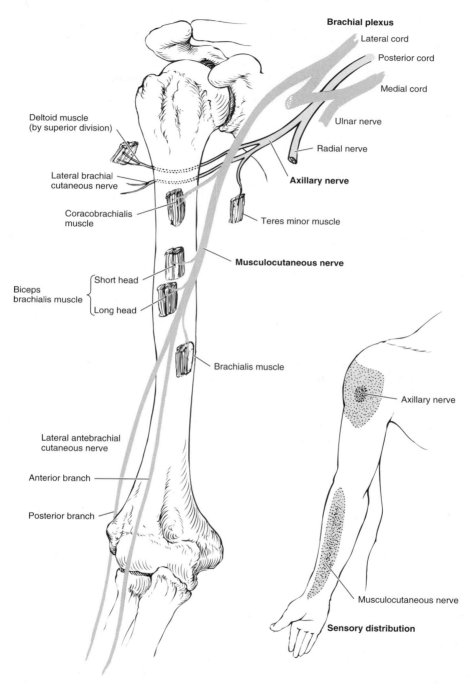

Figure 1-2. The musculocutaneous nerves (C5 and C6) and axillary nerves (C5 and C6). (*Reproduced, with permission, from Waxman SG. Clinical neuroanatomy, 25th ed. New York: McGraw-Hill, 2003:350.*)

the biceps brachii and brachialis muscles. The lateral antebrachial cutaneous nerve to the skin of the lateral forearm represents the terminal continuation of this nerve. Damage to the **musculocutaneous nerve** causes **weakness in supination and flexion of the shoulder and elbow.**

The upper portion of the brachial plexus arises from spinal cord segments C5 and C6; forms the **superior trunk;** and makes major contributions to the **axillary, musculocutaneous, lateral pectoral,** and **suprascapular nerves** and the **nerve to the subclavius muscle.** Injury to the upper plexus typically occurs with an increase in the angle between the shoulder and the neck. This can occur in a newborn during an obstetrical delivery or in adults as the result of a fall on the shoulder and side of the head and neck, which produces a widened angle. The resultant muscle paralysis due to such an injury may be understood more easily in an adult with such an injury. The upper extremity hangs limp by the side because the **deltoid and supraspinatus** (abductors of the arm) are paralyzed as a result of injury of the **axillary** and **suprascapular nerves,** respectively. In addition, the anterior deltoid, biceps brachii, and coracobrachialis (flexors of the arm) are paralyzed due to injury of the **axillary** and **musculocutaneous nerves.** The elbow is extended and the hand is pronated because of paralysis of the **biceps brachii** and **brachialis muscles,** both of which are innervated by the **musculocutaneous nerve.** The extremity is medially rotated because of paralysis of the teres minor and infraspinatus muscles (lateral rotators of the arm) and injury to the axillary and suprascapular nerves. The palm of the hand is turned posteriorly in the "waiter's tip" sign. There is loss of sensation along the lateral aspect of the upper extremity, which corresponds to the **dermatome at C5 and C6.** The upper brachial plexus injury is known as **Erb's** or **Duchenne-Erb palsy.**

The **ulnar nerve (C8 and T1)** is a continuation of the **medial cord,** which enters the posterior compartment through the medial intermuscular septum and passes distally to enter the forearm by curving posteriorly to the **medial epicondyle.** Here it is superficial and at risk for injury. It enters the anterior compartment of the forearm, where it innervates the **flexor carpi ulnaris** and the **bellies of the flexor digitorum profundus** to the **ring and little fingers.** The **ulnar nerve** enters the hand through a **canal (Guyon canal) superficial to the flexor retinaculum.** The nerve supplies all the **intrinsic muscles** of the hand except for the **three thenar muscles** and the **lumbricals** of the **index and middle fingers.** It is sensory to the **medial border of the hand,** the **little finger,** and the **medial aspect of the ring finger.** Damage to the ulnar nerve in the upper forearm causes lateral (radial) deviation of the hand, with weakness in flexion and adduction of the hand at the wrist and loss of flexion at the distal interphalangeal joint of the ring and little fingers. Damage to the ulnar nerve in the upper forearm or at the wrist also results in loss of abduction and adduction of the index, middle, ring, and little fingers due to paralysis of the interossei muscles. A "claw hand" deformity results, and with longstanding damage, **atrophy of the interosseous muscles** occurs.

Injury to the **lower brachial plexus,** known as **Klumpke palsy,** occurs by a similar mechanism, that is, an abnormal widening of the angle between the upper extremity and the thorax. This may occur at obstetrical delivery by traction on the fetal head or when an individual reaches out to interrupt a fall. The roots from **C8 and T1**

and/or the inferior trunk are stretched or torn. Spinal cord segments **C8 and T1** form the **ulnar nerve** and a significant portion of the **median nerve.** Most of the **muscles of the anterior forearm** are **innervated by the median nerve** (see Case 4) and will display weakness. Most of the **muscles of the hand** are innervated by the **ulnar nerve.** There will be loss of sensation along the **median aspect of the arm, forearm, hypothenar eminence,** and **little finger (C8 and T1 dermatome).**

Compression of the brachial plexus cords may occur with prolonged **hyperabduction** during performance of overhead tasks. The **hyperabduction syndrome of pain** down the arm, **paresthesia, hand weakness,** and **skin redness** may result from compression of the cords between the **coracoid process and pectoralis minor.** An **axillary-type crutch** that is too long can compress the posterior cord, leading to radial nerve palsy.

COMPREHENSION QUESTIONS

1.1 A 12-year-old boy is diagnosed with an upper brachial plexus injury after falling from a tree. He presents with his right upper arm lying limp at his side because of loss of abduction. Which of the following muscles are primarily responsible for abduction of the arm at the shoulder?

A. Deltoid and biceps brachii

B. Deltoid and supraspinatus

C. Deltoid and infraspinatus

D. Supraspinatus and infraspinatus

E. Coracobrachialis and supraspinatus

1.2 Injury to the lateral cord of the brachial plexus will also injure its continuation, the musculocutaneous nerve. Which of the following findings would you observe in a patient with this injury?

A. Weakness of abduction of the arm at the shoulder

B. Weakness of adduction of the arm at the shoulder

C. Weakness of extension of the forearm at the elbow

D. Weakness of flexion of the forearm at the elbow

E. Weakness of supination of the forearm and hand

1.3 A 22-year-old man is brought into the emergency department with a knife injury to the axilla. The physician suspects injury to the lower brachial plexus. Which of the following nerves will most likely be affected?

A. Axillary

B. Musculocutaneous

C. Vagus

D. Radial

E. Ulnar

ANSWERS

1.1 **B.** The deltoid and supraspinatus muscles, which are innervated by the axillary and suprascapular nerves, respectively, are the primary abductors of the arm at the shoulder.

1.2 **D.** Injury to the musculocutaneous nerve will result in loss or weakness of flexion at the elbow due to paralysis of the biceps brachii and brachialis muscles.

1.3 **E.** The C8 and T1 portions of the lower brachial plexus make up the majority of the ulnar nerve.

ANATOMY PEARLS

▶ Widening of the angle between the neck and shoulder may stretch the C5 and C6 roots and/or superior trunk, thereby damaging the axillary, musculocutaneous, and suprascapular nerves.

▶ An upper plexus injury results in Erb palsy (or Duchenne-Erb paralysis), which is characterized by an adducted and medially rotated arm, extended elbow, and pronated hand (waiter's tip sign).

▶ The axillary nerve is at risk for fracture of the surgical neck of the humerus.

▶ The musculocutaneous nerve supplies all the muscles of the anterior compartment of the arm.

▶ An abnormal increase in the angle between the upper limb and the thorax and/or severe abduction traction may stretch the C8 and T1 roots and/or the inferior trunk and, hence, affect the ulnar and median nerves.

▶ A lower plexus injury may result in Klumpke palsy, which is characterized primarily by signs of ulnar nerve damage (claw hand).

▶ The ulnar nerve innervates all except five muscles of the hand: the three thenar muscles and the lumbricalis muscles to the index and middle fingers. In ulnar nerve palsies, the patent is unable to abduct and adduct the fingers.

▶ A posterior cord injury results in signs of radial nerve damage (wrist drop).

REFERENCES

Gilroy WM, MacPherson BR, Ross LM. *Atlas of Anatomy*, 2nd ed. New York, NY: Thieme Medical Publishers; 2012:348–349, 352–357.

Moore KL, Dalley AF, Agur AMR. *Clinically Oriented Anatomy*, 7th ed. Baltimore, MD: Lippincott Williams & Wilkins; 2014:704–706, 721–726, 729–730.

Netter FH. *Atlas of Human Anatomy*, 6th ed. Philadelphia, PA: Saunders, 2014: plates 416, 460, 461.

CASE 2

A 32-year-old man is involved in a motor vehicle accident. He used three-point restraints and was driving a sedan. The driver of a pickup truck ran a stop sign while going at approximately 45 mi/h (mph) and broadsided (T-boned) the patient's vehicle on the driver's side. The patient has multiple injuries, including a displaced fracture of the left humerus. He complains of an inability to open his left hand and loss of sensation to a portion of his left hand.

▶ What is the most likely diagnosis?
▶ What is the likely mechanism of the injury?
▶ What portion of the left hand is likely to have sensory deficit?

ANSWERS TO CASE 2:
Radial Nerve Injury

Summary: A 32-year-old man is involved in a motor vehicle accident that causes a displaced fracture of the left humerus. He has motor and sensory losses to his left hand.

- **Most likely diagnosis:** Injury to the radial nerve as it spirals around the humerus, resulting in an inability to extend the wrist or fingers and loss of sensation of the hand

- **Likely mechanism:** Stretch or crush injury to the radial nerve as it spirals around the midshaft of the humerus

- **Likely location of sensory deficit:** Radial (lateral) side of the dorsum of the hand and dorsum of the thumb and index and middle digits

CLINICAL CORRELATION

The radial nerve is at particular risk of injury in its course in the radial groove as it spirals around the midshaft of the humerus. Humeral fractures involving the midshaft region are of particular concern. There is loss of innervation of the posterior extensor muscles in the forearm, resulting in wrist drop and an inability to extend the digits at the metacarpophalangeal joints. The sensory loss on the dorsum of the hand and digits reflects the distal cutaneous distribution of the radial nerve. The triceps muscle (extensor of the elbow) is typically spared; however, fracture-related pain will usually prevent the patient from moving the limb. The deep brachial artery has the same path as the radial nerve in the radial groove and has a similar risk for injury.

APPROACH TO:
The Radial Nerve

OBJECTIVES

1. Be able to describe the origin, course, muscles innervated, and distal cutaneous regions supplied by the radial nerve

2. Be able to describe the arterial blood supply to the upper limb

3. Be able to describe the origin, course, muscles innervated, and distal cutaneous regions supplied by the five major terminal branches of the brachial plexus (see Cases 1, 2, and 4)

DEFINITIONS

FRACTURE: A break in the normal integrity of a bone or cartilage

BLUNT-FORCE TRAUMA: Injury due to a crushing force as opposed to a sharp penetrating force

DISCUSSION

The **radial nerve** is a continuation of the **posterior cord of the brachial plexus**, and it reaches the posterior compartment of the arm by coursing around the **radial groove** of the **humerus** with the **deep brachial artery** (Figure 2-1). It gives off multiple muscular branches to the **triceps muscle** in the posterior compartment. The nerve then pierces the lateral intermuscular septum to return to the anterior compartment of the arm and descends to the level of the lateral epicondyle of the humerus; at this level, it lies deep to the **brachioradialis muscle,** where it divides into its two terminal branches. The **deep branch of the radial nerve is entirely motor to the muscles of the posterior compartment** of the forearm. The other terminal branch, the **superficial branch of the radial nerve,** is sensory to the dorsum of the hand and to the dorsum of the thumb, index finger, and the radial side of the middle finger. The radial nerve also has cutaneous **sensory branches to the posterior and lateral arm and to the posterior forearm.**

The **blood supply to the upper limb** is derived from the **brachial artery**, a direct continuation of the **axillary artery**. It begins at the lower border of the teres major muscle and accompanies the median nerve on the **medial aspect of the humerus**, where its pulsations can be palpated or the artery can be occluded to control hemorrhage. In its descent toward the elbow, it branches off the **deep brachial artery,** which supplies the posterior compartment of the arm, and passes around the **radial groove of the humerus** with the radial nerve. It also has **ulnar collateral branches to the elbow joint**. The **brachial artery** shifts anteriorly as it enters the forearm, lying **just medial to the tendon of the biceps brachii muscle in the cubital fossa.** At about the level of the neck of the radius, it divides into the ulnar and radial arteries, the main arteries of the forearm and hand. Near their origin, each sends recurrent arterial branches to supply the elbow joint.

The **radial artery** supplies the **lateral aspects of the forearm** and at the wrist passes dorsally (deep) through the **anatomical snuffbox** (see Case 3) to become the **deep palmar arch.** The **ulnar artery** is the larger branch of the brachial, and it supplies the **medial aspect of the forearm.** A branch close to its origin, the common interosseous artery, divides into anterior and posterior interosseous arteries. The latter artery is the main blood supply to the posterior compartment. At the wrist, the **ulnar artery** enters the hand to form the **superficial palmar arch**. The **superficial and deep palmar arches** form an **arterial anastomosis** and give rise to **arteries to the digits.**

(See also Case 1.)

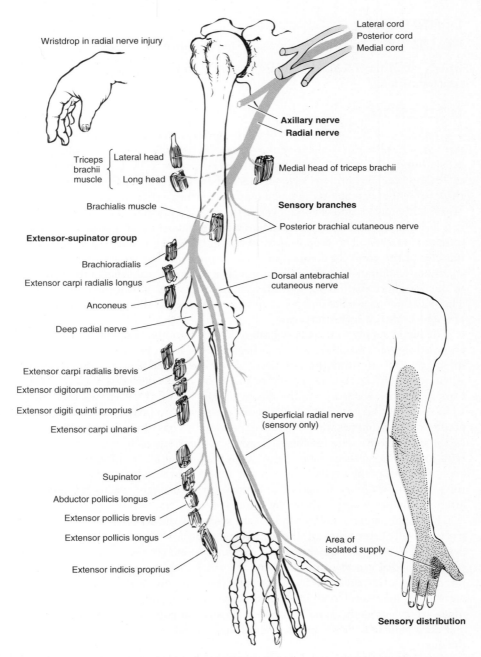

Wristdrop in radial nerve injury

Lateral cord
Posterior cord
Medial cord

Axillary nerve
Radial nerve

Triceps brachii muscle { Lateral head / Long head

Medial head of triceps brachii

Brachialis muscle

Sensory branches

Posterior brachial cutaneous nerve

Extensor-supinator group

Brachioradialis

Extensor carpi radialis longus

Dorsal antebrachial cutaneous nerve

Anconeus

Deep radial nerve

Extensor carpi radialis brevis

Extensor digitorum communis

Extensor digiti quinti proprius

Extensor carpi ulnaris

Superficial radial nerve (sensory only)

Supinator

Abductor pollicis longus

Extensor pollicis brevis

Extensor pollicis longus

Area of isolated supply

Extensor indicis proprius

Sensory distribution

Figure 2-1. The radial nerve. (*Reproduced, with permission, from Waxman SG. Clinical Neuroanatomy, 25th ed. New York: McGraw-Hill, 2003:351.*)

COMPREHENSION QUESTIONS

2.1 An 18-year-old patient has been improperly fitted with axillary-type crutches, which have produced pressure on the posterior cord of the brachial plexus. Which of the following terminal nerves would most likely be affected?
   A. Axillary nerve
   B. Musculocutaneous nerve
   C. Median nerve
   D. Radial nerve
   E. Ulnar nerve

2.2 A 24-year-old man is noted to have a midshaft humeral fracture after falling from a scaffold. Which of the following muscle tests would you perform to test the integrity of the radial nerve?
   A. Flexion of the forearm at the elbow
   B. Flexion of the hand at the wrist
   C. Extension of the hand at the wrist
   D. Abduction of the index, middle, ring, and little fingers
   E. Adduction of the index, middle, ring, and little fingers

2.3 A 45-year-old woman has a severe asthmatic exacerbation and requires an arterial blood gas specimen for management. If you are planning to draw the sample from the brachial artery, where should you insert the needle?
   A. In the lateral aspect of the arm, between the biceps and triceps brachii muscles
   B. Just lateral to the biceps tendon in the cubital fossa
   C. Just medial to the biceps tendon in the cubital fossa
   D. Just medial to the tendon of the flexor carpi radialis muscle at the wrist
   E. Just lateral to the tendon of the flexor carpi ulnaris muscle at the wrist

ANSWERS

2.1 **D.** The radial nerve is a direct continuation of the posterior cord and is affected by injuries to the posterior cord.

2.2 **C.** The radial nerve innervates the muscles of the posterior compartment, which contains the extensors of the wrist.

2.3 **C.** The brachial artery lies superficial and just medial to the tendon of the biceps brachii in the cubital fossa.

ANATOMY PEARLS

▶ The radial nerve supplies all the muscles of the posterior compartment of the arm and forearm. Injury to the radial nerve results in wrist drop.

▶ The brachial artery lies immediately medial to the tendon of the biceps brachii muscle in the cubital fossa.

▶ The superficial and deep palmar arches are formed by the ulnar and radial arteries, respectively.

REFERENCES

Gilroy AM, MacPherson BR, Ross LM. *Atlas of Anatomy*, 2nd ed. New York, NY: Thieme Medical Publishers; 2012:353, 361.

Moore KL, Dalley AF, Agur AMR. *Clinically Oriented Anatomy*, 7th ed. Baltimore, MD: Lippincott Williams & Wilkins, 2014:736, 738, 743, 764, 786.

Netter FH. *Atlas of Human Anatomy*, 6th ed. Philadelphia, PA: Saunders, 2014: plates 418, 465–466.

A 23-year-old male reports that during a game of hoops (basketball), he tripped while driving the ball to the basket, and fell on his outstretched right hand with the palm down. Two days later, he phoned his anatomist father and related that his right wrist was painful. Later that day, he visited his father, who noted that the wrist was slightly swollen and tender but without deformity. He instructed his son to extend the right thumb, thereby accentuating the anatomical "snuffbox," which is extremely tender to deep palpation. His father advised him to get his hand and wrist x-rayed.

▶ What is the most likely diagnosis?
▶ What is the most likely anatomic defect?

ANSWERS TO CASE 3:

Wrist Fracture

Summary: A 23-year-old male trips while playing basketball and suffers trauma to the right wrist. The wrist is slightly swollen, tender, but not deformed. However, deep palpation of the anatomical snuffbox elicits extreme tenderness.

- **Most likely diagnosis:** Wrist fracture
- **Most likely anatomical defect:** Fracture of the narrow middle portion of the scaphoid carpal bone

CLINICAL CORRELATION

This young man tripped while playing basketball and stretched out his right hand to protect himself. His hand, with the palm down and probably deviated to the side of the radius, took the brunt of the fall, resulting in significant impact force to the wrist. This resulted in pain and swelling of the wrist, especially on the radial side, with **point tenderness deep in the anatomical snuffbox.** This is the common mechanism for a **fracture of the scaphoid carpal bone,** the most commonly fractured carpal bone. Point tenderness over a bone or bony process is a hallmark of a fracture at that site. Radiologic confirmation of a fracture is important. The scaphoid bone has a unique blood supply, and proper reduction and alignment of the segments is necessary to decrease the risk of **avascular necrosis.** A fall on an outstretched hand such that it produces hyperextension of the wrist may result in dislocation of the lunate bone. The lunate is usually displaced anteriorly into the carpal tunnel and may impinge on the median nerve. The lunate is the most commonly dislocated carpal bone. A fall on an outstretched palm may also result in a transverse fracture of the distal radius or a **Colles fracture,** which produces a dorsal displacement of the distal fragment, resulting in the characteristic "dinner fork" (also termed "bayonet") deformity. A Smith fracture of the radius in the same region of younger individuals is less common. In a **Smith fracture,** there is trauma to the dorsal aspect of a flexed wrist, and the wrist is deformed with the distal radial fragment displaced ventrally in a "spade" deformity.

APPROACH TO:

The Wrist

OBJECTIVES

1. Be able to describe the bones and joints of the wrist
2. Be able to describe the anatomy of the radius and ulnar as it relates to the transmission of forces in the upper limb and its effect on the forearm bones
3. Be able to describe the boundaries of the anatomical snuff box and its clinical significance

DEFINITIONS

ANATOMICAL SNUFFBOX: Depression on the lateral aspect of the wrist formed by the tendons of the extensor pollicis brevis and abductor pollicis longus anteriorly and the extensor pollicis longus posteriorly

FRACTURE: A break in the normal integrity of a bone or cartilage

AVASCULAR NECROSIS: Death of cells, tissues, or an organ due to insufficient blood supply

DISCUSSION

The junction of the forearm and hand, called the **wrist region,** is a complex of several joints. The articulation of the distal radius with the ulna, called the **distal radioulnar joint,** is the site of movement of the radius anteriorly around the ulna during pronation. The **radius and ulna are united by an articular disk or triangular fibrocartilage** and associated ligaments, which intervenes between the ulna and carpal bones. The **wrist joint proper** is formed between the **distal radius,** the **triangular fibrocartilage,** and the **proximal row of carpal bones.** The **eight carpal bones** are arranged in proximal and distal rows of four bones each. From lateral to medial, the proximal row is composed of the **scaphoid, lunate, triquetrum, pisiform,** and the distal row is composed of the **trapezium, trapezoid, capitate, hamate** (mnemonic: "Some ladies try perfume that they can't handle"). Approximately 50 percent of movement at the wrist occurs at the wrist joint proper, with the remaining 50 percent occurring at the intercarpal joint, between the two rows of carpal bones. A capsule, reinforced by palmar and dorsal radiocarpal ligaments, surrounds the joint. The **radial collateral ligament** strengthens the capsule laterally and limits adduction (ulnar deviation). The **ulnar collateral ligament** strengthens the capsule medially and limits abduction (radial deviation) (Figure 3-1).

In addition to the distal radioulnar joint, the proximal radioulnar joint allows pivotal movement of the radius with the humerus and the ulna during pronation and supination. The **radius and ulna are also joined by the interosseous membrane and its fibers to form a syndesmosis.** The individual fibers are attached proximally on the radius but distally on the ulna. Impact forces on an outstretched hand are transmitted at the wrist to the radius, through the interosseous membrane to the **ulna,** to the **humerus,** and then to the **shoulder,** which is attached to the trunk primarily by muscle. In this fashion, impact forces are transferred distally in the upper limb, with dissipation of the forces as they move proximally. A fall on an outstretched hand may cause fracture of the **radial head** under the right circumstances. Fracture of one forearm bone frequently results in dislocation of the other bone through forces transferred by the **interosseous membrane:**

Wrist → Radius → Interosseous membrane → Ulna → Humerus

The **anatomical snuff box** is bounded anteriorly by the tendons of the **abductor pollicis longus** and the **extensor pollicis brevis** and posteriorly by the **tendon of the extensor pollicis longus.** The **scaphoid bone** and the **radial artery** (a branch of which supplies the scaphoid) lie in the floor of the snuffbox.

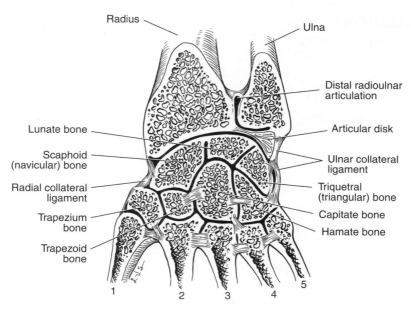

Figure 3-1. Articulations of the bones of the wrist. (*Reproduced, with permission, from Lindner HH. Clinical Anatomy. East Norwalk, CT: Appleton & Lange, 1989:563.*)

COMPREHENSION QUESTIONS

3.1 A 23-year-old accountant trips over a briefcase and falls onto his outstretched hand. A carpal bone fracture is suspected. Which of the following bones is most likely fractured?

    A. Scaphoid

    B. Lunate

    C. Triquetrum

    D. Pisiform

    E. Capitate

3.2 You are examining a radiograph of a patient's wrist and note malalignment (dislocation) of one of the carpal bones. Which of the following is most likely to be the dislocated carpal bone?

    A. Scaphoid

    B. Lunate

    C. Triquetrum

    D. Capitate

    E. Hamate

3.3 A patient with a severe tear of the medial collateral ligament of the wrist would likely display which of the following increased wrist movements?

A. Flexion

B. Extension

C. Abduction

D. Adduction

E. Pronation

3.4 A 24-year-old male slips on a banana peel and falls onto his outstretched hand. Which of the following structures transmits the force from the radius to the ulna?

A. Triangular fibrocartilage

B. Interosseous membrane

C. Scaphoid bone

D. Ulnar collateral ligament

E. Radial collateral ligament

ANSWERS

3.1 **A.** The scaphoid bone is the most frequently fractured carpal bone.

3.2 **B.** The lunate bone is the most frequently dislocated carpal bone.

3.3 **C.** The medial or ulnar collateral ligament limits abduction or radial deviation of the wrist, which would increase if the ligament were severely torn.

3.4 **B.** The interosseous membrane conducts force from the radius to the ulna when the force originates from the wrist.

ANATOMY PEARLS

▶ The union at the distal radioulnar joint is formed by the triangular fibrocartilage.

▶ The chief bony articulation at the wrist is located between the distal head of the radius and the proximal row of carpal bones.

▶ The interosseous membrane forms a fibrous joint between the radius and the ulna, which is important for the transfer and dissipation of impact forces.

▶ The most commonly fractured and dislocated carpal bones are the scaphoid and lunate, respectively.

REFERENCES

Gilroy AM, MacPherson BR, Ross LM. *Atlas of Anatomy*, 2nd ed. New York, NY: Thieme Medical Publishers; 2012:322–327.

Moore KL, Dalley AF, Agur AMR. *Clinically Oriented Anatomy*, 7th ed. Baltimore, MD: Lippincott Williams & Wilkins; 2014:679–680, 686.

Netter FH. *Atlas of Human Anatomy*, 6th ed. Philadelphia, PA: Saunders; 2014: plates 442–444.

A 34-year-old pregnant woman complains of tingling of her right index and middle fingers over a 2-month duration. She notes some weakness of her right hand and has begun to drop items such as her coffee cup. She has otherwise been healthy and denies any trauma or neck pain.

► What is the most likely diagnosis?
► What is the anatomic mechanism for this condition?

ANSWERS TO CASE 4:

Carpal Tunnel Syndrome

Summary: A pregnant woman experiences tingling and weakness of her right index and middle fingers.

- **Most likely diagnosis:** Carpal tunnel syndrome

- **Anatomical mechanism:** Compression of the medial nerve as it passes through the carpal tunnel of the wrist

CLINICAL CORRELATION

The most likely cause for this individual's symptoms is carpal tunnel syndrome. the **carpal tunnel** is a confined, rigid space at the wrist that contains nine tendons with their synovial sheaths and the median nerve. Any condition that further reduces the available space within the tunnel may compress the median nerve, producing numbness and pain in the areas of cutaneous distribution, muscle weakness (especially in the thumb), and muscle atrophy after long-term compression. However, we are not given the distribution of neuropathy of this case. The median nerve may be compressed in several sites along its length between the brachial plexus and the hand, but the carpal tunnel is the most common site. Carpal tunnel syndrome has been associated with endocrine conditions such as diabetes, hypothyroidism, hyperthyroidism, acromegaly, and pregnancy. Other causes include autoimmune disease, lipomas within the canal, hematomas, and carpal bone abnormalities. Females are more commonly affected than males in a ratio of 3:1

Initial treatment is a nighttime splint of the wrist and avoidance of excessive activity with the hand. If symptoms do not decrease, division of the flexor retinaculum (carpal tunnel release) may be necessary.

APPROACH TO:

The Carpal Tunnel

OBJECTIVES

1. Be able to describe the structures that form and pass through the carpal tunnel

2. Be able to describe the course, branches, and muscles innervated by the median nerve in the forearm and hand

3. Be able to describe the skin areas supplied by the median nerve in the hand

4. Be able to describe the course of the ulnar nerve at the wrist as it relates to the carpal tunnel

DEFINITIONS

NEUROPATHY: Any disease or disorder of the peripheral nervous system

CARPAL TUNNEL SYNDROME: Entrapment of the median nerve within the carpal tunnel, resulting in pain, sensory paresthesia, and muscle weakness

MUSCLE ATROPHY: Wasting of muscle tissue, often the result of disuse secondary to interference with its motor innervation

DISCUSSION

The **carpal tunnel** is formed **posteriorly** by the concave surfaces of the carpal bones (see Case 3 for their anatomic arrangement). The **anterior boundary of the tunnel** is formed by a thickening of the deep fascia, the **flexor retinaculum** (transverse carpal ligament). The flexor retinaculum is attached laterally to the tubercles of the scaphoid and trapezium and medially to the **pisiform** and **hook of the hamate.** The carpal tunnel is a passageway for the **nine tendons and their investing synovial sheaths of the flexor muscles of the thumb and fingers:** four tendons each of the **flexor digitorum superficialis** (FDS) and **flexor digitorum profundus** (FDP), the tendon of the **flexor pollicis longus** (FPL), and the **median nerve.** The flexor retinaculum (and the extensor retinaculum dorsally) prevent "bowstringing" of the tendons of the extrinsic hand muscles at the wrist (Figure 4-1).

The **median nerve (C6 through T1)** is formed by contributions from the lateral and medial cords. It passes distally along the arm with the brachial artery and enters the cubital fossa **medial to that artery.** The nerve is at some risk in the **cubital fossa** region. It enters the forearm by passing between the heads of the **pronator teres muscle** and then descends in the forearm between the **FDS** and the **FDP.** In the forearm, the nerve innervates all the **muscles of the anterior compartment** except the flexor carpi ulnaris and the bellies of the FDP to the ring and little fingers. As it approaches the carpal tunnel at the wrist, the **median nerve lies just medial to the tendon of the flexor carpi radialis** muscle and slightly posterior to the tendon of the palmaris longus muscle, if it is present. The **median nerve** enters the hand through the carpal tunnel together with the tendons of the FDS, FDP, and FPL and is at risk for **laceration at the wrist** and compression within the carpal tunnel, deep to the flexor retinaculum (transverse carpal ligament). Typically, the **recurrent branch of the median nerve** arises distal to the flexor retinaculum and the tunnel to innervate the **three thenar muscles: flexor pollicis brevis, abductor pollicis brevis,** and the **opponens pollicis. The lumbrical muscles of the index and middle fingers** receive their motor branches from adjacent common palmar digital branches.

The remainder of the median nerve divides into the **common palmar digital nerves** to the **skin of the thumb and the index, middle, and radial side of the ring fingers,** including their **dorsal nail beds.** The skin of the palm of the hand and thenar eminence is supplied by the palmar cutaneous branch of the median nerve, which typically arises from the median nerve in the distal forearm and does not traverse the carpal tunnel. **Intact skin sensation in the palm** of the hand suggests **carpal tunnel** entrapment of the median nerve, whereas **loss of palmar skin sensation** suggests a **higher nerve lesion.**

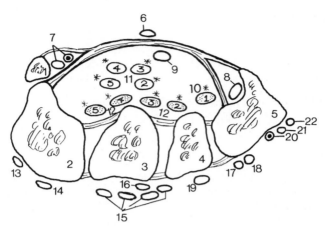

Figure 4-1. Carpal bones in cross section: 1 = pisiform, 2 = hamate, 3 = capitate, 4 = trapezoid, 5 = trapezium, 6 = palmaris longus, 7 = ulnar artery and vein, 8 = flexor carpi radialis, 9 = median nerve, 10 = flexor pollicis longus, 11 = flexor superficialis, 12 = flexor profundus, 13 = extensor carpi ulnaris, 14 = extensor digit minimi, 15 = extensor digitorum, 16 = extensor indicis, 17 = extensor carpi radialis brevis, 18 = extensor carpi radialis, 19 = extensor pollicis longus, 20 = radial artery, 21 = extensor pollicis brevis, 22 = abductor pollicis longus. (*Reproduced, with permission, from the University of Texas Health Science Center in Houston Medical School.*)

Damage to the median nerve in the upper forearm results in loss of pronation, weakness in flexion at the wrist, and medial (ulnar) deviation. There will also be loss of flexion at the proximal interphalangeal joint of the index, middle, ring, and little fingers and loss of flexion at the distal interphalangeal joints of the index and middle fingers. Damage to the median nerve in the upper forearm or at the wrist will also result in loss of flexion, abduction and opposition of the thumb, and flexion at the metacarpal phalangeal joints of the index and middle fingers. Loss of the function of the median nerve results in the "hand of benediction" (a condition in which index and middle fingers are extended, with the ring and small fingers flexed) when the patient is asked to make a fist and an "ape hand" (MP joint extended, PIP and DIP joints flexed) due to longstanding injury with atrophy of the thenar muscles (Figure 4-2).

The ulnar nerve, which innervates all the other intrinsic hand muscles not noted above, enters the hand anterior to the flexor retinaculum and medial to the ulnar artery. The artery and the nerve are covered anteriorly by a condensation of the fascia of the forearm, called the volar carpal ligament. Thus the ulnar nerve and artery come to lie in the Guyon canal, bounded anteriorly by the volar carpal ligament, posteriorly by the flexor retinaculum, medially by the pisiform, and laterally by the hook of the hamate.

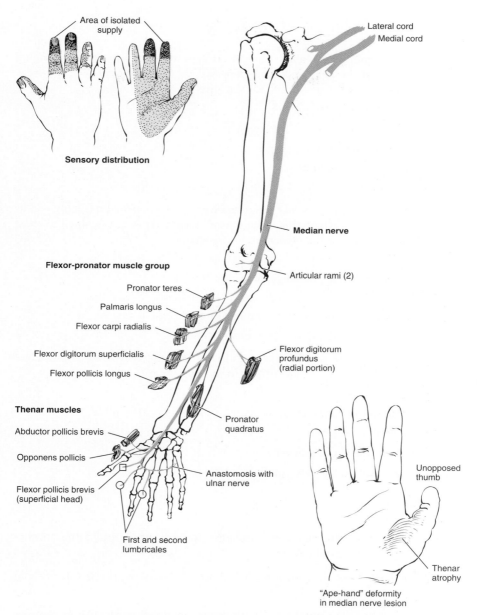

Figure 4-2. The median nerve. (*Reproduced, with permission, from Waxman SG. Clinical Neuroanatomy, 25th ed. New York: McGraw-Hill, 2003:352.*)

COMPREHENSION QUESTIONS

4.1 You are examining an axial (cross-sectional) magnetic resonance imaging (MRI) scan of the wrist and have identified the carpal tunnel. Which of the following is the structure that forms the anterior wall of the tunnel?

A. Palmar aponeurosis

B. Volar carpal ligament

C. Flexor retinaculum

D. Extensor retinaculum

E. Deep fascia

4.2 As you are explaining carpal tunnel syndrome to a woman who has the condition, you show her where the median nerve is located just as it is about to enter the tunnel. Where is the median nerve located?

A. Just lateral to the flexor carpi radialis tendon

B. Just medial to the flexor carpi radialis tendon

C. Just medial to the flexor palmaris longus tendon

D. Just lateral to the flexor carpi ulnaris tendon

E. Just medial to the flexor carpi ulnaris tendon

4.3 If the median nerve were severed in an industrial accident at the wrist, which of the following muscles would still retain their function?

A. Flexor pollicis brevis

B. Abductor pollicis brevis

C. Opponens pollicis

D. Lumbricals of the index and middle fingers

E. Lumbricals of the ring and little fingers

ANSWERS

4.1 **C.** The flexor retinaculum or transverse carpal ligament forms the anterior boundary of the carpal tunnel.

4.2 **B.** The median nerve lies just medial to the tendon of the flexor carpi radialis at the wrist.

4.3 **E.** The lumbricals to the ring and little finger are innervated by the ulnar nerve.

ANATOMY PEARLS

▶ All muscles of the anterior compartment of the forearm are supplied by the median nerve except for the flexor carpi ulnaris and the medial half of the flexor digitorum profundus, which are innervated by the ulnar nerve.

▶ Injury to the median nerve results in the hand of benediction when one attempts to make a fist.

▶ The carpal tunnel is formed by the flexor retinaculum and the eight carpal bones.

▶ The carpal tunnel contains nine tendons (four for the FDS, four for the FDP, and one for the FPL) and the median nerve.

▶ The median nerve supplies five muscles in the hand (flexor pollicis brevis, abductor pollicis brevis, opponens pollicis, lumbricals 1 and 2); the skin of the thumb; and the index, middle, and lateral ring fingers.

▶ The palmar cutaneous branch of the median nerve does not traverse the carpal tunnel.

▶ The ulnar nerve does not traverse the carpal tunnel; it enters the hand anterior to the flexor retinaculum in Guyon canal.

REFERENCES

Gilroy AM, MacPherson BR, Ross LM. *Atlas of Anatomy*, 2nd ed. New York, NY: Thieme Medical Publishers; 2012:370–371.

Moore KL, Dalley AF, Agur AMR. *Clinically Oriented Anatomy*, 7th ed. Baltimore, MD: Lippincott Williams & Wilkins, 2014:761–764, 786, 790–792.

Netter FH. *Atlas of Human Anatomy*, 6th ed. Philadelphia, PA: Saunders, 2014: plates 447, 449–450.

While playing football, a 17-year-old defensive end was attempting to tackle a full-back with an outstretched left arm. The arm was hit with substantial force, and he now complains of severe shoulder pain and his left arm is hanging down with some external rotation. The pain prevents him from moving the limb. A radiograph is negative for a fracture, but the head of the humerus is superimposed on the neck of the scapula.

► What is the most likely diagnosis?
► What is the most likely nerve injured?

ANSWERS TO CASE 5:
Shoulder Dislocation

Summary: A 17-year-old football player's left arm was outstretched and hit with some force. He has shoulder pain, and his arm hangs limp down his side with external rotation. There is no fracture, and the humeral head is superimposed on the scapular neck.

- **Most likely diagnosis:** Glenohumeral joint dislocation (shoulder dislocation)

- **Most likely nerve injured:** Axillary nerve

CLINICAL CORRELATION

The shoulder is the most commonly dislocated large joint of the body and is usually dislocated in an anterior direction. Typically, the dislocation is also inferior such that the humeral head is located inferior and lateral to the coracoid process. The humeral head will often have an infraglenoid and infraclavicular position. The diagnosis may be difficult to make. The typical mechanism consists in a violent force to the humerus that is abducted and externally rotated, resulting in extension of the joint; this action displaces the humeral head inferiorly, thus tearing the weak inferior portion of the shoulder joint capsule. This is facilitated by the fulcrum effect of the acromion. The strong flexor and adductor muscles pull the humeral head anteriorly and medially to the usual subcoracoid position. Typically, the patient will not move the arm and will support the limb flexed at the elbow with the opposite hand. The arm will be slightly abducted and medially rotated. The usually rounded curve of the shoulder is lost, and there is a depression evident inferior to the acromion. The humeral head is palpable, if not visible, in the deltopectoral triangle. First priorities are assessment of the neural and vascular integrity of the upper limb by testing motor and sensory functions of the fingers and palpation of the radial pulse. Different methods to reduce the dislocation exist, including the modified **Hippocratic method,** in which one operator pulls on a sheet placed around the thorax of the patient, while a second operator gently applies traction on the wrist of the affected side. Other injuries that may accompany a shoulder dislocation include strain on the tendons of the subscapularis and supraspinatus muscles, tears of the glenoid labrum, fracture of the greater tubercle of the humerus, trauma to the axillary nerve (as demonstrated by loss of sensation in the shoulder patch region over the deltoid muscle), and trauma to the axillary artery or its branches, such as the posterior circumflex humeral or subscapular arteries.

APPROACH TO:
The Shoulder

OBJECTIVES

1. Be able to describe the bones and joints that make up the shoulder girdle

2. Be able to delineate the anatomy of the glenohumeral joint

3. Be able to list the extrinsic muscles of the shoulder, their action at the shoulder, and their innervation

4. Be able to describe the components of the rotator cuff and their action, innervation, and functional importance to the shoulder.

DEFINITIONS

SHOULDER: Junction between the arm and the trunk.

SHOULDER GIRDLE: The clavicle, the scapula, and the proximal humerus.

DISCUSSION

The **shoulder girdle** and the **shoulder joint** proper consist of the **clavicle,** the **scapula,** and the **proximal portion of the humerus.** The only bony articulation between the shoulder girdle and the trunk occurs at the sternoclavicular joint. This strong joint has two joint spaces created by a cartilage articular disk. The synovial articulation of the clavicle with the manubrium of the sternum is strengthened by a joint capsule, anterior and posterior sternoclavicular, and interclavicular and costoclavicular ligaments. The lateral end of the **clavicle** articulates with the **acromion of the scapula** to form the **acromioclavicular joint.** An incomplete articular disk is present within this synovial joint. A thin, loose capsule surrounds the acromioclavicular joint, which is reinforced superiorly by an **acromioclavicular ligament,** but its chief strength and support is derived from the **trapezoid and conoid ligaments,** which together form the **coracoclavicular ligament.**

The articulation of the **glenoid cavity** on the neck of the scapula with the head of the humerus forms the **glenohumeral joint.** This **shallow ball-and-socket synovial joint** forms the shoulder joint proper. The anatomy of this joint allows a **wide range of motion,** although **stability is decreased.** The diameter of the humeral head is about three times greater than the diameter of the glenoid cavity, which is increased somewhat by a rim of fibrocartilage attached to the margin of the glenoid, the **glenoid labrum.** The joint capsule attaches to the margin of the glenoid proximally and to the anatomical neck of the humerus distally. The **capsule has openings** for the tendon of the **long head of the biceps muscle** and **for the subscapular bursa,** which communicates with the joint cavity. **Three glenohumeral ligaments,** bandlike thickenings of the anterior capsule, are identifiable only internally (Figure 5-1). The **coracohumeral ligament** reinforces the capsule superiorly, and the **transverse humeral ligament** bridges the **intertubercular groove** with the tendon and synovial sheath

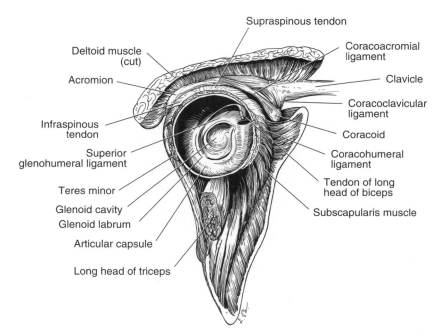

Supraspinous tendon

Deltoid muscle (cut)

Coracoacromial ligament

Acromion

Clavicle

Infraspinous tendon

Coracoclavicular ligament

Coracoid

Superior glenohumeral ligament

Coracohumeral ligament

Teres minor

Tendon of long head of biceps

Glenoid cavity

Subscapularis muscle

Glenoid labrum

Articular capsule

Long head of triceps

Figure 5-1. Anatomy of the shoulder joint. (*Reproduced, with permission, from Lindner HH. Clinical Anatomy. East Norwalk, CT: Appleton & Lange, 1989:528.*)

of the long head of the biceps brachii muscle. The **roof of the glenohumeral joint** is formed by the inferior surface of the **acromion** and **the coracoacromial ligament.**

The upper limb is attached to the trunk primarily by muscles. This group of muscles, the extrinsic muscles of the shoulder, originates from the trunk and inserts onto the scapula in most instances or the humerus directly. The action of muscles attaching to the scapula produces movement of the scapula, which greatly increases the range of motion at the shoulder. The extrinsic muscles and the action and innervation of each are listed in Table 5-1.

Table 5-1 • EXTRINSIC MUSCLES OF THE SHOULDER		
Muscle	**Action**	**Innervation**
Trapezius	Retracts, elevates, depresses, rotates scapula	Spinal accessory nerve
Latissimus dorsi	Extends, adducts, medially rotates arm	Thoracodorsal nerve
Levator scapulae	Elevates, rotates scapula	Dorsal scapular nerve
Rhomboid major and minor	Retracts and rotates scapula	Dorsal scapular nerve
Serratus anterior	Protracts and rotates scapula	Long thoracic nerve
Pectoralis major	Adducts and medially rotates arm	Lateral and medial pectoral nerves
Pectoralis minor	Stabilizes scapula	Medial pectoral nerve

Table 5-2 • INTRINSIC MUSCLES OF THE SHOULDER

Muscle	Action	Innervation
Deltoid	Abducts, flexes, and extends arm	Axillary nerve
Teres major	Adducts and medially rotates arm	Lower subscapular nerve
Supraspinatus*	Initiates abduction of arm	Suprascapular nerve
Infraspinatus*	Laterally rotates arm	Suprascapular nerve
Teres minor*	Laterally rotates arm	Axillary nerve
Subscapularis*	Adducts and medially rotates arm	Upper and lower subscapular nerve

*Rotator cuff muscles.

The **intrinsic muscles of the shoulder** originate from the scapula and insert onto the humerus. They include the **deltoid,** the **teres major,** and the **rotator cuff muscles.** The **rotator cuff tendons** surround and blend with the capsule of the glenohumeral joint and provide **major strength and stability to the joint.** The intrinsic muscles of the shoulder and their actions and innervations are presented in Table 5-2. The tendon of the **supraspinatus muscle** passes **superior** to the capsule, between it and the acromion and deltoid muscle to insert onto the **greater tubercle.** The **subacromial (subdeltoid) bursa** intervenes between the tendon and the undersurface of the acromion and the deltoid muscle. Nevertheless, the **supraspinatus tendon** is typically damaged with rotator cuff tears.

COMPREHENSION QUESTIONS

5.1 You are evaluating a radiograph of the only bony articulation between the upper limb and the trunk. Which of the following joints are you evaluating?

A. Glenohumeral
B. Acromioclavicular
C. Humeroclavicular
D. Coracoclavicular
E. Sternoclavicular

5.2 You are explaining the anatomy of the shoulder to a young athlete who has sustained an injury to one of his shoulders. You tell him that the chief stability to this joint is from which of the following?

A. Glenohumeral ligaments
B. Acromioclavicular ligament
C. Rotator cuff muscles
D. Coracoclavicular ligaments
E. Coracohumeral ligament

5.3 A college baseball pitcher has shoulder discomfort, and you suspect a rotator cuff tear. You will most likely see damage to the tendon of which of the following muscles?

A. Supraspinatus

B. Infraspinatus

C. Subscapularis

D. Teres major

E. Teres minor

ANSWERS

5.1 **E.** The upper limb is attached to the trunk only at the sternoclavicular joint. The primary attachment is muscular.

5.2 **C.** The primary stability to the glenohumeral joint is provided by the tendons of the rotator cuff.

5.3 **A.** The tendon of the supraspinatus is typically damaged in a rotator cuff tear due to the narrow space between the head of the humerus and the acromion.

ANATOMY PEARLS

▶ Shoulder dislocations are common, are almost always anterior, and place the axillary nerve at risk.

▶ The only bony articulation between the upper limb and the trunk is at the sternoclavicular joint. The primary attachment of the limb to the trunk is by musculature.

▶ The shallow ball-and-socket glenohumeral joint permits a wide range of motion, but with decreased stability.

▶ The tendons of the rotator cuff muscles provide primary strength and stability to the glenohumeral joint.

▶ The supraspinatus muscle tendon blends with the superior capsule. Although it is protected from the undersurface of the acromion by the subacromial (subdeltoid) bursa, its tendon is usually injured in rotator cuff tears.

REFERENCES

Gilroy AM, MacPherson BR, Ross LM. *Atlas of Anatomy*, 2nd ed. New York, NY: Thieme Medical Publishers; 2012:282–287.

Moore KL, Dalley AF, Agur AMR. *Clinically Oriented Anatomy*, 7th ed. Baltimore, MD: Lippincott Williams & Wilkins, 2014:704–707, 712, 796–800, 814–815.

Netter FH. *Atlas of Human Anatomy*, 6th ed. Philadelphia, PA: Saunders, 2014: plates 405–408.

CASE 6

A 37-year-old male accountant is picked up by his wife at his office. He enters the passenger seat of their automobile and turns to fasten the seatbelt as his wife begins to exit the parking lot. Another vehicle entering the lot strikes their vehicle head on, and he is thrown forward by the sudden deceleration. His left knee strikes the dashboard violently, and he feels a painful pop in his left hip. After ambulance transport to the hospital emergency department, he is noted to experience severe pain in the left hip region. His left lower limb is noted to be adducted and medially rotated and shorter than his right lower limb There is a painful mass in the lateral gluteal region.

- ▶ What is the most likely diagnosis?
- ▶ What structures are likely to be involved in this injury?
- ▶ What clinically important structures are at potential risk?

ANSWERS TO CASE 6:

Posterior Hip Dislocation

Summary: A 37-year-old male automobile passenger was turning to his right during a head-on collision, in which his left knee struck the dashboard. He experiences a painful pop in his left hip. His left hip is very painful, and his left lower limb is shortened, adducted, and medially rotated. A large painful mass is present in the lateral gluteal area.

- **Most likely diagnosis:** Posterior hip dislocation, with or without acetabular fracture

- **Structures involved:** Hip joint, including the femoral head, joint capsule and ligaments, and acetabulum

- **Structures at risk:** Sciatic nerve

CLINICAL CORRELATION

This male automobile passenger has suffered a deceleration injury in which his left knee has forcefully impacted the vehicle's dashboard, forcing the femoral head posteriorly over the rim of the acetabulum. As he turned to reach for the seatbelt, his hip was flexed because he was in sitting position, and was adducted and medially rotated, the classic hip position for this type of injury. The painful pop was the femoral head tearing through the posterior joint capsule and ligaments. The painful lateral gluteal mass is the femoral head on the lateral aspect of the ilium, and the affected limb appears shorter than the other because of the abnormal position of the femoral head. The first step would be radiographic confirmation of the dislocation or dislocation fracture. This would be followed by urgent reduction of the dislocation to decrease the risk of avascular necrosis of the femoral head and other complications such as posttraumatic arthritis. Fracture repair, if a fracture is present, may be performed at a later time.

APPROACH TO:

The Hip Joint

OBJECTIVES

1. Be able to describe the anatomy of the hip joint, including the proximal femur, the joint capsule and ligaments, and the acetabulum

2. Be able to describe the course of the sciatic nerve as it relates to the hip joint

DEFINITIONS

HIP BONE: Os coxae, irregular flat bone formed by the fusion of the pubis, ilium, and ischium

HIP JOINT: Joint formed by the acetabulum and the head of the femur

SCIATIC NERVE: The largest nerve of the body, formed from the sacral plexus by the ventral rami of L4 through S3, anatomically uniting the common fibular and tibial nerves

DISCUSSION

The **hip joint** is a unique ball-and-socket joint formed by the femur and hip bone that combines stability for the transference of weight to the lower limbs and enables an erect posture, with a wide range of motion. The **ball portion** of the joint is formed by the **head of the femur** and is angled **superomedially** to the shaft by the neck. The large adjacent **greater and lesser trochanters** serve as points of attachment for muscles acting across the hip joint. The **socket** portion of the joint is formed by the **cup-shaped acetabulum** on the lateral surface of the hip bone (os coxae). The hip bone is formed by the fused **ilium, ischium,** and **pubic bones,** all of which participate in forming the **acetabulum.** The depth of the acetabulum is increased by a **C-shaped rim** of **fibrocartilage** called the **acetabular labrum.** The incomplete inferior portion of the labrum, the acetabular notch, is completed by the **transverse acetabular ligament,** and the weak, intraarticular **ligament of the femoral head** passes from the head to the acetabulum adjacent to the notch. The femoral head and the acetabulum are covered with articular cartilage.

The **hip joint** is surrounded by a **capsule,** lined with a synovial membrane, and strengthened by three ligamentous thickenings, named for their proximal attachments. The capsule is reinforced anteriorly and superiorly by the strong, **inverted Y-shaped iliofemoral ligament.** Inferior and posterior capsular thickenings are the **pubofemoral and ischiofemoral ligaments,** respectively. The blood supply to the hip joint originates from the **lateral and medial circumflex arteries,** typically branches of the **deep femoral artery;** the **medial circumflex artery** is the most important. These **vessels** reach the femoral head along the neck, where they are at risk for **femoral neck fractures.** A small artery of the head (a branch of the obturator artery) runs within the ligament of the head.

In addition to producing movement at the hip joint, the **muscles** crossing the joint impart much of the **joint's stability** while the subject stands erect. The movements at the hip joint and the muscles (with their innervation) producing these movements are listed in Table 6-1

The **large sciatic nerve** (Figure 6-1) exits the pelvis through the **greater sciatic foramen** and enters the **deep gluteal region** immediately **inferior to the piriformis muscle.** The sciatic nerve is actually the **combined common fibular nerve (lateral portion)** and the **tibial nerve (medial portion).** The common fibular nerve innervates the lateral compartment muscles (superficial fibular branch) and the anterior compartment muscles (deep fibular branch) of the leg. **The tibial portion innervates the muscles of posterior compartments of the thigh and leg (calf) and the sole of the foot.** The sciatic nerve lies **posterior to the hip joint.**

Table 6-1 • MUSCLES ACTING ON THE HIP JOINT

	Muscles	Innervation
Flexion	Iliopsoas (psoas)	Anterior rami L1–L3 nerves
	Iliopsoas (iliacus), rectus femoris, pectineus, sartorius	Femoral nerve
	Adductor longus, brevis, and magnus, gracilis	Obturator nerve
	Tensor fascia lata	Superior gluteal nerve
Extension	Hamstrings: semitendinosus, semimembranosus, long head of biceps femoris, adductor magnus (hamstring portion)	Tibial portion of sciatic nerve
	Gluteus maximus	Inferior gluteal nerve
Abduction	Gluteus medius and minimus, tensor fascia lata	Superior gluteal nerve
Adduction	Adductor longus, brevis and magnus, gracilis, obturator externus	Obturator nerve
	Pectineus	Femoral nerve
Medial rotation	Gluteus medius and minimus, tensor fascia lata	Inferior gluteal nerve
Lateral rotation	Obturator internus, piriformis, superior and inferior gemelli, quadratus femoris	Direct nerves to muscles arising from L5, S1–S2
	Obturator externus	Obturator nerve
	Gluteus maximus	Inferior gluteal nerve

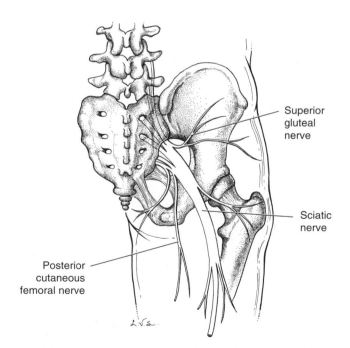

Figure 6-1. The sciatic nerve. (*Reproduced, with permission, from Lindner HH. Clinical Anatomy. East Norwalk, CT: Appleton & Lange, 1989:591.*)

COMPREHENSION QUESTIONS

6.1 In a patient who has a posterior dislocation of the hip, which of the following ligamentous structures would be torn?

A. Pubofemoral ligament

B. Iliofemoral ligament

C. Ischiofemoral ligament

D. Lacunar ligament

E. Sacrotuberous ligament

6.2 A 54-year-old man has just dislocated his right hip. The physician is concerned about the integrity of the joint's blood supply. Which artery supplies most of the blood supply to the hip joint?

A. Lateral circumflex femoral

B. Medial circumflex femoral

C. Superficial circumflex iliac

D. Deep circumflex iliac

E. Perforating

6.3 A patient with a hip dislocation is also exhibiting weakness of extension of the thigh at the hip. This would indicate possible damage to which of the following?

A. Femoral nerve

B. Obturator nerve

C. Common fibular portion of the sciatic nerve

D. Tibial portion of the sciatic nerve

E. Saphenous nerve

ANSWERS

6.1 **C.** Posterior dislocation of the hip would tear the ischiofemoral ligament, thus reinforcing the capsule of the hip posteriorly.

6.2 **B.** The chief blood supply to the hip is the medial circumflex femoral artery.

6.3 **D.** Most of the extensor muscles of the hip (the hamstrings) are innervated by the tibial portion of the sciatic nerve. The gluteus maximus muscle, innervated by the inferior gluteal nerve, could still weakly extend the thigh at the hip.

ANATOMY PEARLS

▶ The bones forming the hip bone, ilium, ischium, and pubis converge to form the acetabulum.

▶ The strongest ligament reinforcing the hip capsule is the iliofemoral or Y ligament.

▶ The most important artery supplying the hip joint is the medial circumflex femoral artery.

REFERENCES

Gilroy AM, MacPherson BR, Ross LM. *Atlas of Anatomy*, 2nd ed. New York, NY: Thieme Medical Publishers; 2012:386–389.

Moore KL, Dalley AF, Agur AMR. *Clinically Oriented Anatomy*, 7th ed. Baltimore, MD: Lippincott Williams & Wilkins, 2014:626–634, 660–661.

Netter FH. *Atlas of Human Anatomy*, 6th ed. Philadelphia, PA: Saunders, 2014: plates 473–475, 490–491.

CASE 7

A 25-year-old woman is on her first ski trip to Colorado. She has advanced from the bunny (gentle) slopes and, during the last run of the day, falls and twists her right leg. She cannot stand on her right leg because of pain and is brought down the hill in a snowmobile. On examination, the right knee is swollen and tender. With the patient sitting on the stretcher with her knee flexed, the lower leg seems to have several centimeters of excess anterior mobility.

▶ What is the most likely diagnosis?
▶ What is the mechanism of the injury?

ANSWER TO CASE 7:

Anterior Cruciate Ligament Rupture

Summary: A 25-year-old woman twists her right lower limb in a ski accident. She has right knee swelling and tenderness, and excessive anterior mobility with the knee flexed.

- **Most likely diagnosis:** Anterior cruciate ligament (ACL) tear
- **Mechanism of injury:** Excessive rotational force strains or ruptures the ligament

CLINICAL CORRELATION

Injuries to the knee are very common because it bears weight, combines mobility in flexion and extension, and allows some rotation. The stability of the knee depends entirely on its ligaments and muscles. Sports injuries to the knee are most commonly caused by high-speed and rotational forces applied to the leg through the knee joint. In addition, certain ligaments are anatomically related to the menisci, on which the distal femur articulates. This 25-year-old woman was involved in a ski injury, a common setting for ACL injury. The twisting force to the lower limb when a ski becomes lodged in snow and the body continues to rotate can produce significant trauma to the knee. The ACL passes from the posterior aspect of the distal femur to the intercondylar region of the anterior aspect of the proximal tibia; it limits anterior movement of the tibia in relation to the femur. Thus, on examination, this patient exhibits the "anterior drawer sign," or excessive anterior mobility of the tibia with the knee flexed. This injury will usually require surgical repair.

APPROACH TO:

The Knee Joint

OBJECTIVES

1. Be able to describe the anatomy of the knee joint, including the bones, ligaments, possible movements, and muscles responsible for these movements

2. Be able to describe the mechanism of injury to the four main ligaments of the knee

DEFINITIONS

KNEE: Hinge joint between the femur and proximal tibia

PATELLA: A triangular bone approximately 5 cm in diameter situated in the front of the knee at the insertion of the quadriceps tendons

MENISCUS: Crescent-shaped intraarticular cartilage

DISCUSSION

The **knee joint** is a **synovial hinge joint** formed by the distal **femur**, the **proximal tibia**, and the **patella.** It is a relatively stable joint; its movements consist primarily of flexion and extension, with some gliding, rolling, and locking rotation. The **distal femur** forms two large knuckle-like **lateral and medial condyles**, which articulate with **lateral and medial tibial condyles.** The superior surfaces of the tibial condyles are flattened to form the **tibial plateau.** An **intercondylar eminence** fits between the femoral condyles, and the proximal fibula articulates with the lateral tibial condyle but is not a part of the knee joint. The **patella** articulates with the femur anteriorly. The flat tibial condylar surfaces are modified to accommodate the femoral condyles by the **C-shaped lateral and medial menisci.** These fibrocartilaginous structures are wedge-shaped in cross section; they are thick peripherally but thin internally, are firmly attached to the tibial condyles, and serve as shock absorbers. The **lateral meniscus** is the smaller of the two, and is somewhat **circular**, whereas the **medial meniscus is C-shaped.** The femoral and remaining portions of the tibial condyles are covered with articular cartilage (Figure 7-1).

The knee joint is surrounded by a capsule, lined with synovial membrane, and reinforced by several ligamentous thickenings. Anteriorly, the **patella is embedded within the tendon of the quadriceps femoris muscle group.** Inferior to the patella, the tendon becomes the patellar ligament, which inserts into the **tibial tuberosity.** Laterally, the capsule is thickened to form the **fibular (lateral) collateral ligament** from the lateral femoral epicondyle to the fibular head. The fibular collateral ligament remains separated from the lateral meniscus by the tendon of the popliteus muscle. It prevents increase of the lateral angle or adduction of the leg at the knee. The **tibial (medial) collateral ligament** extends from the medial femoral epicondyle to the medial tibial condyle. The deep aspect of this ligament is firmly attached to the margin of the medial meniscus. It prevents increase of the medial angle or abduction of the leg at the knee. Posteriorly, the capsule is reinforced by **oblique and arcuate popliteal ligaments.** The knee is unique because of the presence of two intraarticular ligaments: **the ACL and the posterior cruciate ligament (PCL).** The cruciate ligaments are covered by synovial membrane and thus are external to the synovial cavity and are named for their attachment to the tibia. The ACL extends from the anterior tibial plateau near the intercondylar eminence to the posteromedial aspect of the lateral femoral condyle. It **limits anterior displacement of the tibial in relation to the femur** and limits hyperextension. The **PCL** extends from the posterior aspect of the tibial plateau to the anterolateral aspect of the medial femoral condyle. In its course, it crosses the ACL on its medial side and is **larger and stronger** than the ACL. The **PCL limits posterior displacement** of the tibia on the femur and limits hyperflexion. A dozen or so **bursae** are associated with the knee joint, and four of these communicate with the synovial cavity of the joint: suprapatellar, popliteus, anserine, and gastrocnemius. Thus, inflammation of any of these bursae (bursitis) will likely result in swelling of the entire knee joint. The knee joint is richly supplied by several **genicular and recurrent arteries from the femoral, popliteal, and anterior tibial arteries.**

Additional strength and stability to the knee joint are provided by the muscles that cross and produce movement at the joint. The action and innervation of these muscles are listed in Table 7-1.

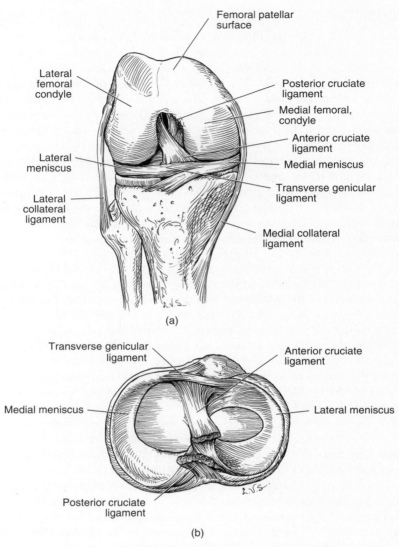

Femoral patellar
surface

Lateral
femoral
condyle

Posterior cruciate
ligament

Medial femoral,
condyle

Anterior cruciate
ligament

Lateral
meniscus

Medial meniscus

Transverse genicular
ligament

Lateral
collateral
ligament

Medial collateral
ligament

(a)

Transverse genicular
ligament

Anterior cruciate
ligament

Medial meniscus

Lateral meniscus

Posterior cruciate
ligament

(b)

Figure 7-1. The knee joint. Ligaments of the knee in full extension (a). The superior aspect of the knee showing the menisci (b). (*Reproduced, with permission, from Lindner HH. Clinical Anatomy. East Norwalk, CT: Appleton & Lange, 1989:615.*)

An abnormal **force applied to the lateral aspect of the knee** with the foot planted **stretches the tibial (medial) collateral ligament,** causing a sprain or, if sufficiently forceful, rupture of this ligament. The exposed lateral knee makes this injury more frequent. Because the medial meniscus is firmly attached to the deep surface of the tibial collateral ligament, it also is frequently damaged. Forces applied to the medial aspect of the knee can damage the fibular (lateral) collateral ligament in a similar manner. However, because the lateral meniscus is not attached to the ligament,

Table 7-1 • MUSCLES ACTING ON THE KNEE JOINT		
	Muscles	**Innervation**
Extension	Quadriceps group: rectus femoris, vastus lateralis, interomedialis, and medialis	Femoral nerve
Flexion	Hamstring group: semitendinosus, semimembranosus, and biceps femoris	Tibial portion of sciatic nerve
Lateral rotation	Biceps femoris	Tibial portion of sciatic nerve
Medial rotation	Popliteus, semitendinosus, and semimembranosus	Tibial nerve

it typically is not damaged. Excessive force to the anterior aspect of the tibia will move it posteriorly, thus stretching or tearing the PCL. The **ACL** is most often damaged when forces or activities produce **hyperextension of the knee.**

COMPREHENSION QUESTIONS

7.1 Your patient has sustained an external force to the knee. Which of the following ligaments has prevented abduction of the leg at the knee?

A. Oblique popliteal

B. Anterior cruciate

C. Posterior cruciate

D. Lateral collateral

E. Medial collateral

7.2 In this same patient, which of the following ligaments prevented posterior displacement of the tibia on the femur?

A. Oblique popliteal

B. Anterior cruciate

C. Posterior cruciate

D. Lateral collateral

E. Medial collateral

7.3 You have examined a patient and find that there is weakness in the ability to flex the knee. This indicates a problem with which of the following nerves?

A. Femoral nerve

B. Tibial nerve

C. Common fibular nerve

D. Deep fibular nerve

E. Superficial fibular nerve

ANSWERS

7.1 **E.** Abduction of the leg at the knee is limited by the medial or tibial collateral ligament.

7.2 **C.** Posterior displacement of the tibia on the femur is limited by the PCL.

7.3 **B.** The muscles that flex the knee are innervated by the tibial portion of the sciatic nerve.

ANATOMY PEARLS

► The **anterior** cruciate ligament is so named because it is attached to the **anterior** aspect of the tibia and prevents **anterior** displacement of the tibia on the femur.

► The tibial collateral ligament is attached to the medial meniscus; thus, both are often injured by an abnormal force to the lateral knee.

► The ACL is injured most often by hyperextension of the knee.

REFERENCES

Gilroy RM, MacPherson BR, Ross LM. *Atlas of Anatomy*, 2nd ed. New York, NY: Thieme Medical Publishers; 2012:406–414.

Moore KL, Dalley AF, Agur AMR. *Clinically Oriented Anatomy*, 7th ed. Baltimore, MD: Lippincott Williams & Wilkins; 2014:634–645, 662–663.

Netter FH. *Atlas of Human Anatomy*, 6th ed. Philadelphia, PA: Saunders; 2014: plates 494–498.

A 23-year-old female is seen on the postpartum floor the day after delivering a 9-lb baby boy. She is concerned about her right foot, which has become numb and weak since delivering the baby. Walking has been difficult for her because her right foot tends to drop, and her toes drag. When asked about her labor course, she reports that she had an epidural with satisfactory pain relief but a difficult and prolonged pushing stage of labor (3 h) in stirrups. She denies any back pain or problems with the other leg. On exam, she has decreased sensation on the top of the right foot and lateral side of the lower leg along with an inability to dorsiflex the right foot, resulting in a foot drop. Minimal peripheral edema is seen in both lower extremities.

▶ What is the most likely diagnosis?
▶ What factors likely led to this condition?

ANSWER TO CASE 8:

Common Fibular Nerve Injury

Summary: A 23-year-old female postpartum day 1 with right foot weakness, numbness, and foot drop after a difficult vaginal delivery.

- **Most likely diagnosis:** Common fibular nerve injury (compression)

- **Factors leading to injury:** Prolonged compression of common fibular nerve by stirrups and flexion at the knee

CLINICAL CORRELATION

Compression of the common fibular nerve during labor is the most common postpartum nerve injury of the lower extremity. Compression of the common fibular nerve occurs from both the flexion of the knees and compression from stirrup on lateral aspect of knee. The common fibular nerve may also be injured during knee surgery, trauma, or prolonged periods of compression (coma, deep sleep, lower-extremity cast). Because of the epidural anesthesia, the patient likely felt no pain from the prolonged compression. Injury to the common fibular nerve causes numbness, weakness to the lower leg and foot, and foot drop (inability to dorsiflex the foot). The majority of compression injuries after delivery are self-limiting and improve with supportive care. Proper positioning of the patient requires a good understanding of anatomy to avoid periods of prolonged nerve compression.

APPROACH TO:

Lower Limb

OBJECTIVES

1. Be able to describe the origin, course, muscles innervated, and distal cutaneous regions supplied by the sciatic nerve and its tibial and common fibular branches

2. Be able to describe the origin, course, muscles innervated, and distal cutaneous regions supplied by the femoral and obturator nerves

DEFINITIONS

EPIDURAL: The space external to the spinal cord's dura mater; anesthetic agents are injected into this space for epidural anesthesia

NERVE COMPRESSION: Pressure on a nerve such that neural transmission is temporarily blocked

DORSIFLEXION: Decrease in the angle between the lower leg and foot, as with walking on one's heels; the opposite of plantarflexion, as standing on tiptoe

DISCUSSION

The **sciatic nerve** (L4–S3) is the largest nerve in the body, arising from the lumbo-sacral plexus. It exits the pelvis through the **greater sciatic foramen**, inferior to the piriformis muscle (see Figure 8-1). The sciatic nerve is actually two nerves, the **tibial (medial) and common fibula (lateral) nerves,** loosely bound together by connective tissue. The tibial nerve is derived from the anterior division of the anterior rami, while the common fibular is derived from the posterior division of the anterior rami. No muscles of the gluteal region are innervated by the sciatic nerve. It descends in the posterior compartment of the thigh, where its **tibial nerve innervates all the muscles (hip extensors and knee flexors) of the posterior thigh** except the short head of the biceps femoris (innervated by the common fibular nerve). At approximately superior angle of the popliteal fossa, the tibial and common fibular portions separate.

The **common fibular nerve** passes laterally, superficially, and **courses around the neck of the fibula** subcutaneously, **where it risks injury or compression** (Figure 8-2). It then divides into its superficial fibular nerve, which innervates the fibular muscles (evertors) of the lateral compartment of the leg, and the skin of the lateral leg and dorsum of the foot. The deep fibular nerve enters the anterior compartment of the leg, and innervates the muscles of this compartment (dorsiflexors), intrinsic dorsal foot muscles, and skin between the great and second toe. **Severing the deep fibular nerve results in a foot drop.**

The **tibial nerve** descends through the popliteal fossa and enters the posterior compartment of the leg to innervate the posterior compartment muscles (plantarflexors and invertors). It also gives off a medial sural branch which joins the communicating sural branch of the common fibular to form the **sural nerve,** which is sensory to the posterior aspect of the leg and lateral foot. At the level of the posterior malleolus, the tibial nerve divides into the **lateral and medial plantar nerves,** which innervate the intrinsic muscles and skin of the sole of the foot. **Severing the tibial nerve in the leg results in an inability to stand on tiptoe.**

The **femoral nerve** (L2–L4) arises from the lumbar plexus. It exits the abdomen posterior to the inguinal ligament and lies lateral to and outside the femoral sheath and its contents. It innervates the muscles (hip flexors and knee extensors) of the anterior compartment of the thigh and the skin of the anterior thigh and medial leg. The **obturator nerve** (L2–L4) exits the abdomen through the obturator canal and enters the medial compartment of the thigh to innervate these muscles (adductors) and a patch of skin on the medial side of the thigh.

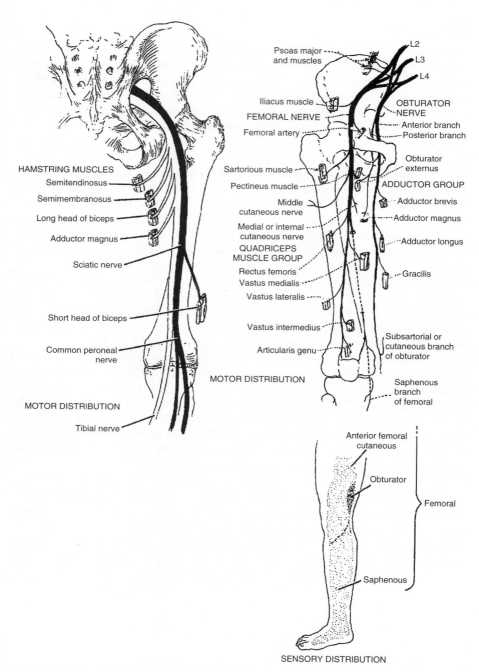

Figure 8-1. Innervation of the thigh. (*Reproduced, with permission, from Lindner HH. Clinical Anatomy. Norwalk, CT: Appleton & Lange, 1989.*)

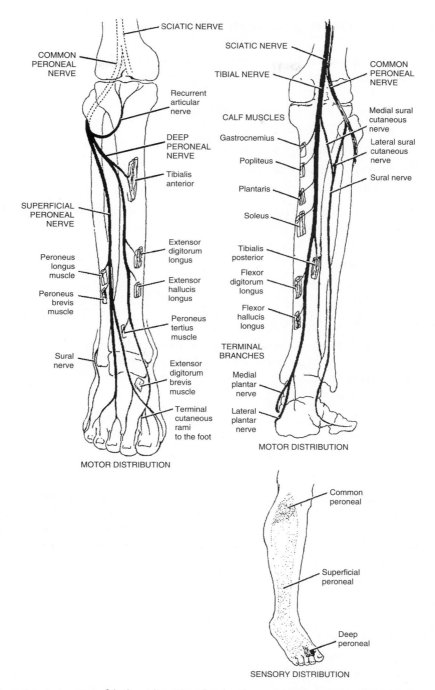

Figure 8-2. Innervation of the lower leg. (*Reproduced, with permission, from Lindner HH. Clinical Anatomy. Norwalk, CT: Appleton & Lange, 1989:50.*)

COMPREHENSION QUESTIONS

8.1 During an abdominal hysterectomy for a cancerous uterus, the obturator nerve was accidentally severed. This resulted in the patient losing which of the following actions?

A. Extension of the leg at the knee

B. Extension of the thigh at the hip

C. Adduction of the thigh at the hip

D. Flexion of the leg at the knee

E. Dorsiflexion of the foot at the ankle

8.2 A patient comes to you complaining of his inability to stand on tiptoe. Which of the following nerve injuries is most likely to be involved?

A. Femoral nerve

B. Tibial nerve

C. Common fibular nerve

D. Deep fibular nerve

E. Superficial fibular nerve

8.3 A 32-year-old woman is brought into the emergency department because she is unable to evert her foot at the ankle. Which of the following nerve injuries is most likely to be involved?

A. Femoral nerve

B. Obturator nerve

C. Tibial nerve

D. Deep fibular nerve

E. Superficial fibular nerve

8.4 A 14-year-old male is placed in a left lower leg cast after a skateboarding accident. After he has been in the cast for 3 weeks, he complains of some numbness of the "top of the left foot." On exam, he is noted to be unable to dorsiflex his left foot. What is the most likely location of the nerve compression in this patient?

A. Lateral malleolus

B. Medial malleolus

C. Tarsal canal

D. Fibular head

E. Popliteal fossa

ANSWERS

8.1 **C.** The obturator nerve innervates the muscles of the medial compartment of the thigh which adduct the thigh at the hip.

8.2 **B.** The plantarflexors are located in the posterior compartment of the leg and innervated by the tibial nerve.

8.3 **E.** The muscles of the lateral compartment of the leg evert the foot and are innervated by the superficial fibular nerve.

8.4 **D.** This young man likely has compression of the common peroneal nerve as the nerve traverses laterally around the fibular head, where it is relatively superficial and not well protected. Injury to the common peroneal nerve leads to "foot drop" and inability to dorsiflex.

ANATOMY PEARLS

▶ The muscles of the posterior thigh, leg, and sole of the foot are all innervated by the tibial nerve (except the short head of the biceps femoris).

▶ The dorsiflexor muscles are innervated by the deep fibular nerve.

REFERENCES

Gilroy AM, MacPherson BR, Ross LM. *Atlas of Anatomy*, 2nd ed. New York, NY: Thieme Medical Publishers; 2012:446–448.

Moore KL, Dalley AF, Agur AMR. *Clinically Oriented Anatomy*, 7th ed. Baltimore, MD: Lippincott Williams & Wilkins; 2014:574–575, 586–587, 592, 596.

Netter FH. *Atlas of Human Anatomy*, 6th ed. Philadelphia, PA: Saunders; 2014: plates 525–529.

CASE 9

A 42-year-old diabetic woman complains of soreness of the left leg. She is moderately obese and has been recovering from surgical removal of her gallbladder (cholecystectomy) performed 2 weeks ago. On examination, she has obvious swelling in the left lower leg and some tenderness of the calf that increases when the calf is gently squeezed. There is no redness of the leg, and she is afebrile (without fever).

▶ What is the most likely diagnosis?
▶ What structure is likely affected?

ANSWER TO CASE 9:

Deep Venous Thrombosis

Summary: A 42-year-old obese diabetic woman complains of soreness of the left leg. She had gallbladder surgery 2 weeks previously. Her left calf is tender but without erythema. She is afebrile.

- **Most likely diagnosis:** Deep venous thrombosis (DVT)

- **Structure likely affected:** Anterior and posterior tibial veins and fibular veins

CLINICAL CORRELATION

Venous thrombosis, or pathological blood clots within a vein, is a common cause of morbidity and mortality. The Virchow triad of venous stasis, hypercoagulability, and vessel wall damage comprise notable risk factors. This patient has several risk factors for DVT. She is obese, diabetic, and has been inactive because of postoperative bedrest, with the latter producing venous stasis. Although gynecological and orthopedic surgeries especially predispose individuals to DVT, any surgery increases the risk. Prevention of DVT includes using lower-limb compression devices during and after surgery. These devices intermittently squeeze the legs, thereby simulating the muscular contraction of physical activity. Anticoagulant therapy, such as small-dose heparin, is also sometimes given before surgery and for 1 or 2 days postoperatively. If DVT is confirmed with ultrasound or radiologically with venous contrast (venogram), anticoagulation therapy is important to decrease the risk of embolization of the thrombosis, which can travel directly to the lungs to produce potentially fatal pulmonary embolism.

APPROACH TO:

Vascular Supply of Lower Extremity

OBJECTIVES

1. Be able to draw the arterial blood supply to the lower limb

2. Be able to describe the deep and superficial venous drainage of the lower limb

DEFINITIONS

AFEBRILE: Without fever

EMBOLUS: A mass such as part of the blood clot (thrombus), air, or fat that travels through a vessel and lodges and obstructs blood flow

THROMBOSIS: Process by which a blood clot forms within a blood vessel

DISCUSSION

The **chief blood supply to the lower limb is from the femoral artery**, the continuation of the **external iliac artery** inferior to the inguinal ligament, within the **femoral triangle**. The femoral triangle is bounded by the **inguinal ligament superiorly**, the **sartorius muscle laterally**, and the **adductor longus muscle medially**. It contains the femoral nerve and the femoral sheath and its contents. The femoral artery lies in the lateral compartment of the **femoral sheath**, with the **femoral vein** medial to it, and the femoral canal with its associated inguinal lymph nodes medial to the vein. The femoral nerve lies lateral to and outside the femoral sheath. Just inferior to the inguinal ligament, the **superficial epigastric, superficial circumflex iliac**, and **two external pudendal arteries** arise from the femoral artery. Within the **femoral triangle**, the deep femoral artery arises and descends posteriorly to the femoral vessels and the adductor longus muscle. The **lateral and medial circumflex arteries** usually arise from the **deep femoral artery**, as do muscular branches and several perforating branches, to supply the posterior thigh. As the femoral artery descends toward the apex of the femoral triangle, it enters the **adductor canal** and becomes the **popliteal artery**, and it assumes a position posterior to the femur. It descends inferiorly through the popliteal fossa, giving rise to five genicular arteries to the knee, and terminates by dividing into the **anterior and posterior tibial arteries** near the lower border of the popliteus muscle (Figure 9-1).

The **anterior tibial artery** pierces the **interosseus membrane**, from which it descends through the **anterior compartment**, supplying structures in this compartment, and terminates anterior to the ankle by becoming the **dorsal artery of the foot**. The dorsal artery and its lateral tarsal branch form an **arch of the dorsum of the foot** and provides the chief blood supply to the toes. The **posterior tibial artery** descends in the posterior compartment and supplies it and the lateral compartment by perforating branches in addition to its fibular branch. It passes posteriorly to the **medial malleolus**, enters the sole of the foot, and divides into lateral and medial plantar arteries that supply the sole of the foot.

Other arteries that supply portions of the lower limb include the **obturator artery**, which supplies the medial compartment of the thigh. The **superior and inferior gluteal and the internal pudendal arteries** provide the chief blood supply to the gluteal region.

The lower limb has **superficial and deep systems of veins,** both of which terminate in the **femoral vein,** which continues superiorly to the inguinal ligament as the external iliac vein. The deep system of veins usually consists of **paired venae comitantes**, which accompany the arteries for which they are named. Thus anterior and posterior tibial veins are formed from the dorsum and sole of the foot. Fibular veins arise in the posterior compartment and drain blood to the posterior tibial veins, which ascend and are joined by the anterior tibial veins to form the popliteal vein. The **popliteal vein** becomes the **femoral vein** as it traverses the **adductor canal**, receives the deep femoral vein in the femoral sheath, and enters the abdomen beneath the inguinal canal to become the external iliac vein. A deep vein of the thigh accompanies its artery and drains into the femoral vein. The superficial system of veins is composed of the **small and great saphenous veins** and is found in the superficial fascial of the limb. The small saphenous is formed posterior to the lateral

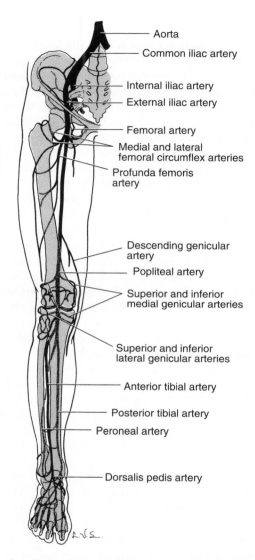

Aorta
Common iliac artery
Internal iliac artery
External iliac artery
Femoral artery
Medial and lateral
femoral circumflex arteries
Profunda femoris
artery
Descending genicular
artery
Popliteal artery
Superior and inferior
medial genicular arteries
Superior and inferior
lateral genicular arteries
Anterior tibial artery
Posterior tibial artery
Peroneal artery
Dorsalis pedis artery

Figure 9-1. Arterial supply to the leg. (*Reproduced, with permission, from Lindner HH. Clinical Anatomy. East Norwalk, CT: Appleton & Lange, 1989:602.*)

malleolus and ascends in the middle of the calf to terminate in the popliteal vein in the popliteal fossa. The **great saphenous vein is formed from the dorsal venous arch of the foot anterior to the medial malleolus.** It ascends along the **medial aspect of the leg and the thigh.** It pierces the saphenous opening in the fascia lata (deep fascia of the thigh) to empty into the femoral vein within the femoral sheath. Numerous communications exist between the two saphenous veins. Of greater clinical importance, **communications between the superficial and deep systems exist as perforating branches** whose valves are arranged to allow venous flow from superficial to deep, but not in the opposite direction. This important shunt allows muscular contraction to produce venous return against the effects of gravity.

COMPREHENSION QUESTIONS

9.1 While operating on the posterior compartment of the thigh, an orthopedic surgeon takes care to preserve the arterial blood supply to the muscles in that region. These are branches of which of the following arteries?

A. Deep femoral artery

B. Femoral artery

C. Superior gluteal artery

D. Inferior gluteal artery

E. Obturator artery

9.2 A patient has sustained lower-limb trauma that has damaged the posterior tibial artery. Therefore, you will be concerned about the blood supply to which of the following?

A. Posterior thigh only

B. Lateral compartment of the leg only

C. Posterior compartment of the leg only

D. Sole of the foot only

E. Posterior compartment of the leg and the sole of the foot

9.3 Which are the chief deep veins of the leg that are of concern for DVT?

A. Small saphenous vein

B. Great saphenous vein

C. Deep femoral vein

D. Anterior and posterior tibial veins

E. Obturator vein

ANSWERS

9.1 **A.** The blood supply to the posterior compartment of the thigh originates from perforating branches of the deep femoral artery.

9.2 **E.** The posterior tibial artery provides the blood supply to the calf and the sole of the foot.

9.3 **D.** The deep veins of the leg are the anterior and posterior tibial veins that accompany the arteries of the same name.

ANATOMY PEARLS

▶ The relationship between lateral medial of structures within the femoral triangle is defined by the acronym NAVeL (femoral Nerve, Artery, Vein; empty space, Lymph nodes).

▶ The chief blood supply to the thigh and the hip arises from the deep femoral artery.

▶ The posterior tibial artery enters the foot through the tarsal canal, posterior to the medial malleolus.

▶ Venous blood flow is from superficial to deep venous systems.

REFERENCES

Gilroy AM, MacPherson BR, Ross LM. *Atlas of Anatomy*, 2nd ed. New York NY: Thieme Medical Publishers; 2012:446–447.

Moore KL, Dalley AF, Agur AMR. *Clinically Oriented Anatomy*, 7th ed. Baltimore, MD: Lippincott Williams & Wilkins; 2014:532–535, 540, 551–556, 602–603.

Netter FH. *Atlas of Human Anatomy*, 6th ed. Philadelphia, PA: Saunders; 2014: plates 470–471, 487, 499, 505.

A 42-year-old man is brought to the emergency room complaining of intense pain in his left calf and ankle. He had entered a tennis tournament with his 15-year-old son and states that, as he lunged after a hard-hit serve, he heard a "snap," fell to the court in tremendous pain, and could not walk. On examination, the left calf is tender and indurated, with an irregular mass noted in the back of the midcalf area.

► What is the most likely diagnosis?
► What type of excessive abnormal ankle movement would be present?

ANSWER TO CASE 10:

Achilles Tendon Rupture

Summary: A 42-year-old man heard a "snap" while playing tennis and experienced left calf pain after lunging for a ball. The left calf is tender and indurated and has a lump.

- **Most likely diagnosis:** Achilles tendon rupture
- **Likely abnormal ankle movement present:** Dorsiflexion

CLINICAL CORRELATION

The gastrocnemius and soleus muscles form a three-headed muscle group (triceps surae) that unite to form a single tendon, the calcaneal or Achilles tendon, which inserts into the calcaneus bone. These muscles produce plantar flexion of the foot at the ankle and limit dorsiflexion. Running or quick-start athletic activity, such as described in this case, may lead to strain or rupture of the tendon. The snap heard by this patient is fairly common in calcaneal tendon avulsion. The mass noted in the left calf is due to foreshortening of the triceps surae. Compared with the opposite side, the affected foot will have greater range of motion in dorsiflexion and loss of plantar flexion. Treatment is usually surgical repair of the tendon. Because of the limited blood supply to this tendon, a long immobilization is typically required. Postoperative physical therapy to prevent tendon contracture is critical.

APPROACH TO:

The Ankle Joint

OBJECTIVES

1. Be able to describe the anatomy of the ankle joint
2. Be able to describe the muscles that cross the ankle joint, the movements they produce, and the ligaments that limit these movements

DEFINITIONS

INDURATED: Process in which usually soft tissue becomes extremely hard

STRAIN: Injury that results from overuse or inappropriate use

AVULSION: Violent separation or tearing away

DISCUSSION

Movements of the foot at the ankle occur at two joints: the **ankle joint proper** or **talocrural joint**, which is formed by the distal ends or malleoli of the fibula and tibia, and the **trochlea of the talus bone**. A **mortise-shaped joint** is formed at which

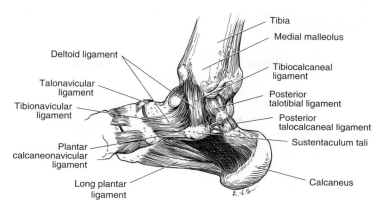

Figure 10-1. The medial ligaments of the ankle joint. (*Reproduced, with permission, from Lindner HH. Clinical Anatomy. East Norwalk, CT: Appleton & Lange, 1989:638.*)

the hinged movements of **dorsi- and plantarflexion** occur. The ankle joint is more stable in dorsiflexion because the anterior aspect of the trochlea is tightly wedged between the lateral and medial malleoli. The movements of **inversion and eversion** of the foot occur primarily at the **subtalar joint** (between the talus and calcaneus bones), but also at the transverse tarsal joint with articulation of the talus and calcaneus bones with the navicular and cuboid bones (Figures 10-1 and 10-2).

The capsule of the ankle joint is thin anteriorly and posteriorly, but ligaments reinforce the capsule laterally and medially to provide much of the stability. A **relatively**

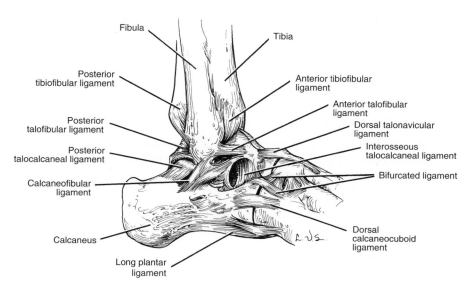

Figure 10-2. The lateral ligaments of the ankle joint. (*Reproduced, with permission, from Lindner HH. Clinical Anatomy. East Norwalk, CT: Appleton & Lange, 1989:639.*)

Table 10-1 • MUSCLES ACTING ON THE FOOT		
	Muscle	Innervation
Dorsiflexion	Tibialis anterior, extensor digitorum longus, extensor hallucis longus, fibularis tertius	Deep fibular nerve
Plantarflexion	Triceps surae: gastrocnemius, soleus, plantaris; flexor hallucis longus, flexor digitorum longus, tibialis posterior	Tibial nerve
Inversion	Tibialis anterior	Deep fibular nerve
	Tibialis posterior	Tibial nerve
Eversion	Fibularis longus, and brevis	Superficial fibular nerve
	Fibularis tertius	Deep fibular nerve

weak lateral ligament is formed by **three individual ligaments,** all of which attach to the lateral malleolus of the fibula: anterior and posterior talofibular ligaments and calcaneofibular ligaments. The **lateral ligament limits excessive inversion.** The medial (deltoid) ligament is a very strong ligament composed of four individual ligaments that attach to the tibia: tibionavicular, anterior and posterior tibiotalar, and tibiocalcaneal ligaments. The **medial ligament limits eversion.** The muscles that produce dorsiflexion at the ankle are located in the anterior compartment of the leg, whereas the muscles that cause plantar flexion and eversion are located in the posterior and lateral compartments, respectively. The muscles that produce movements of the foot at the ankle are listed in Table 10-1.

COMPREHENSION QUESTIONS

10.1 When will a patient's ankle joint have the greatest stability?
    A. When the knee is flexed
    B. When the foot is dorsiflexed
    C. When the foot is plantarflexed
    D. When the foot is everted
    E. When the foot is inverted

10.2 You are concerned that your patient's medial deltoid ligament may have been torn from its proximal attachment. Which of the following would you palpate for tenderness?
    A. The medial aspect of the tibial shaft
    B. The lateral aspect of the fibular shaft
    C. The lateral malleolus
    D. The medial malleolus
    E. The calcaneus

10.3 Your female patient is unable to walk on her tiptoes. You immediately suspect damage to which of the following nerves?

A. Sural nerve

B. Tibial nerve

C. Common fibular nerve

D. Superficial fibular nerve

E. Deep fibular nerve

ANSWERS

10.1 **B.** The talocrural or ankle joint proper has the greatest stability in dorsiflexion.

10.2 **D.** The four components of the deltoid ligament arise from the medial malleolus.

10.3 **B.** Plantarflexion of the foot at the ankle is produced by the muscles in the calf, which are innervated by the tibial nerve.

ANATOMY PEARLS

▶ Dorsiflexion and plantarflexion occur at the ankle joint proper, whereas inversion and eversion occur primarily at the subtalar joint.

▶ A patient with a lesion of the tibial nerve above the knee would be unable to stand on tiptoe (plantarflex the foot at the ankle).

▶ A patient with foot drop and inability to evert the foot (walk on the instep) has a lesion of the common fibular nerve (which is at risk as it passes around the neck of the fibula).

REFERENCES

Gilroy AM, MacPherson BR, Ross LM. *Atlas of Anatomy*, 2nd ed. New York, NY: Thieme Medical Publishers; 2012:418, 422, 435, 439, 467.

Moore KL, Dalley AF, Agur AMR. *Clinically Oriented Anatomy*, 7th ed. Baltimore, MD: Lippincott Williams & Wilkins; 2014:596–600, 607, 647–650.

Netter FH. *Atlas of Human Anatomy*, 6th ed. Philadelphia, PA: Saunders, 2014: plates 504, 506, 514.

A 48-year-old man complains of swelling of the neck and shortness of breath of 1-week duration. He has noticed some nasal stuffiness with hoarseness of his voice for about 3 weeks and had attributed these symptoms to an upper respiratory infection. He denies the use of alcohol but has smoked two packs of cigarettes per day for 30 years. Lately, he feels as though something is pushing against his throat. On physical examination, the patient's face appears ruddy and swollen. The jugular veins are distended.

▶ What is the most likely diagnosis?
▶ What is the most likely cause?
▶ What are the anatomical structures involved?

ANSWER TO CASE 11:
Superior Vena Cava Syndrome

Summary: A 48-year-old heavy smoker has a 1-week history of neck swelling, dyspnea, and the sensation of something pushing on his throat. Three weeks ago, he developed nasal stuffiness and voice hoarseness. He has facial plethora, edema, and jugular venous distention.

- **Most likely diagnosis:** Superior vena cava (SVC) syndrome

- **Most likely cause:** Bronchogenic lung cancer

- **Anatomical structures likely involved:** SVC, trachea, and right mainstem bronchus

CLINICAL CORRELATION

The SVC receives venous drainage from the head, neck, upper limb, and thorax. Located in the upper mediastinum, this thin-walled vessel is susceptible to pressure from external sources. The most common cause of such external compression is malignancy, usually from a right-side bronchogenic carcinoma. Such tumors can also compress the trachea, producing dyspnea, and may involve the recurrent laryngeal nerve, producing hoarseness, as in this patient. The stellate sympathetic ganglion may be compressed, leading to Horner syndrome, the clinical triad of unilateral miosis (constricted pupil), facial anhydrosis (dryness), and ptosis (drooping eyelid). The development of SVC syndrome is often an emergency because the trachea may be obstructed, leading to respiratory compromise. The priority in treatment is directed toward the airway, with oxygen and possibly diuretic agents, and corticosteroid agents to relieve the edema. A chest radiograph, computed tomographic (CT) scan, and a tissue biopsy, in that order, would be the next diagnostic steps. Most patients who have lung cancer are treated with radiotherapy. Although patients who have SVC syndrome often respond well to the radiation treatment, the overall prognosis is nearly always poor due to the advanced extent of the cancer.

APPROACH TO:
The Mediastinum

OBJECTIVES

1. Be able to describe the divisions of the mediastinum and the contents of each

2. Be able to describe the lymphatic drainage of the thoracic organs

DEFINITIONS

SUPERIOR VENA CAVA SYNDROME: Engorgement of the vessels of the head, neck, and upper limbs accompanied by cough and respiratory difficulty due to compression of the SVC or its main tributaries by a benign or malignant mass.

BRONCHOGENIC CARCINOMA: A malignant tumor arising from the mucosal epithelium of the large bronchi.

MEDIASTINUM: The central region of the thorax between the two pleural cavities.

DISCUSSION

The **mediastinum** is the central portion of the thoracic cavity, and it lies between the two pulmonary cavities. It is bounded laterally by the mediastinal pleura. It contains all the thoracic viscera except the two lungs. **Superior and inferior divisions** are described, with the latter further divided into **anterior, middle, and posterior divisions.**

The **superior mediastinum** extends from the superior thoracic aperture bounded by the superior border of the manubrium, first rib, and T1 vertebral body. The inferior boundary is a horizontal line from the sternal angle posterior to the intervertebral disk between T4 and T5. The superior mediastinum contains the following structures, from anterior to posterior: adipose tissue with remnants of the thymus gland, right and left brachiocephalic vein, SVC, aorta with its brachiocephalic trunk, left common carotid and left subclavian arterial branches, trachea, esophagus, and thoracic duct. Related to these structures are the phrenic, vagus, left recurrent laryngeal and cardiac nerves, and anterior mediastinal lymph node group (Figure 11-1).

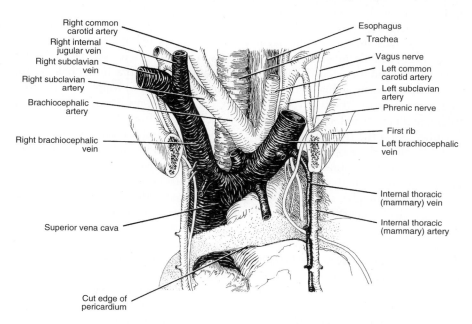

Figure 11-1. The superior mediastinum and root of the neck. (*Reproduced, with permission, from Lindner HH. Clinical Anatomy. East Norwalk, CT: Appleton & Lange, 1989:226.*)

The **inferior mediastinum** is bounded anteriorly by the sternum, posteriorly by vertebral bodies T5 through T12, and the diaphragm inferiorly. The **anterior mediastinum** portion lies between the sternum and the pericardial sac and contains small branches of the internal thoracic artery and a few nodes of the parasternal lymph node group. The thymus gland is present during childhood. The **middle mediastinum** contains the pericardial sac with the heart, terminations of the SVC, inferior vena cava (IVC), pulmonary veins, the ascending aorta, the pulmonary trunk and its bifurcations into the right and left pulmonary arteries, lung roots, phrenic nerve, and bronchial lymph nodes. The **posterior mediastinum** lies between the pericardial sac and vertebral bodies T5 through T12. It contains the esophagus, descending thoracic aorta and right intercostals and esophageal arteries, azygous venous system, thoracic duct, vagus and splanchnic nerves, and posterior mediastinal lymph nodes.

The body's main lymphatic vessel, the **thoracic duct**, originates in the abdomen at the level of L1 as a highly variable dilation called the **cisterna chili.** It enters the posterior mediastinum through the aortic hiatus and lies on the right anterior surface of the thoracic vertebral bodies, posterior to the esophagus between the **azygous venous system** and the thoracic aorta. By the level of the sternal angle, the duct completes a shift to the left side, traverses the superior mediastinum, and terminates by emptying into the venous system near the junction of the left internal jugular and subclavian veins. The **thoracic duct receives lymph drainage from the lower limbs, abdomen and left hemithorax, upper limb, and head and neck.** A small right lymphatic duct receives lymph drainage from the right hemithorax, upper limb, and head and neck. The thoracic and right lymphatic ducts are described as receiving lymph from jugular, subclavian, and bronchomediastinal trunks, although these trunks may variably unite or empty into veins independently.

Lymph node groups that drain lymph from the thoracic wall include the parasternal, intercostals, and several diaphragmatic groups. Lymph nodes that drain thoracic viscera include anterior mediastinal nodes in the anterior region of the superior mediastinum and those located on the anterior surfaces of the brachiocephalic veins, SVC, and the aortic arch and its branches. These nodes receive lymph from the thymus, inferior part of the thyroid gland, heart, pericardium, mediastinal pleura, lung hilum, and parasternal and diaphragmatic nodes. Vessels from the anterior mediastinal help form the right and left bronchomediastinal trunks. Posterior mediastinal nodes lie along the esophagus and thoracic aorta and drain lymph from the esophagus, pericardium, diaphragm, and superior surface of the liver. The vessels from this group empty into the thoracic duct or tracheobronchial nodes (Figure 11-2).

The largest number of visceral nodes are associated with the lungs and airways. The lungs have **superficial and deep lymphatic plexuses** that drain into the bronchopulmonary (hilar) lymph node. The **deep plexuses,** however, first drain through pulmonary nodes along the bronchi within the lung, from which the lymph passes to the bronchopulmonary nodes. Lymph then drains to **inferior and superior tracheobronchial nodes** (below and above the tracheal bifurcation) and **tracheal nodes** located along the sides of the trachea. The tracheobronchial nodes on the right side are closely related to the SVC and receive lymph from the right lung and the inferior part of the left lung. Vessels from these node groups form the right and left bronchomediastinal trunks.

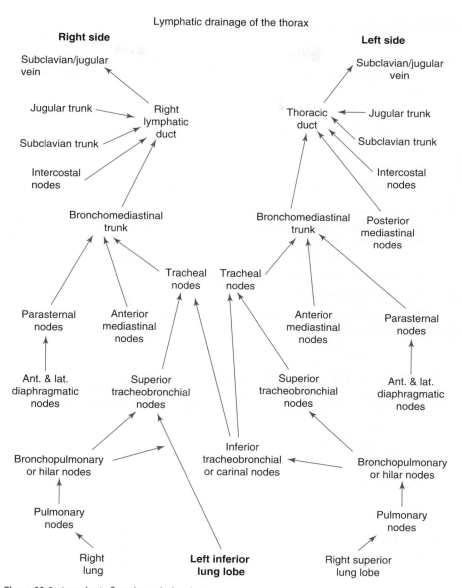

Figure 11-2. Lymphatic flow through the chest.

COMPREHENSION QUESTIONS

11.1 A thoracic surgeon has entered the right pleural cavity and excised two suspicious lymph nodes at the hilum of the right lung for frozen-section pathological study. These nodes belong to which of the following lymph node groups?

A. Parasternal

B. Paratracheal

C. Superior tracheobronchial

D. Inferior tracheobronchial

E. Bronchopulmonary

11.2 During a surgical procedure, a surgeon has reflected the fat pad containing the thymic remnants and notes a large venous structure crossing the midline from the left and apparently emptying into the SVC. This vessel is most likely which of the following?

A. Right brachiocephalic vein

B. Left brachiocephalic vein

C. Left internal jugular vein

D. Left subclavian vein

E. Azygous vein

11.3 A pediatric heart surgeon has just divided the sternum in a child to repair a cardiac malformation. A lobulated gland-like structure is seen immediately obscuring the heart. This is most likely which of the following?

A. Lung

B. Thyroid gland

C. Thymus

D. Lymph nodes

E. Liver

ANSWERS

11.1 **E.** The bronchopulmonary lymph node group is located at the hilum of each lung, and it receives lymph from the superficial and deep lymphatic plexuses.

11.2 **B.** The left brachiocephalic vein crosses the midline to unite with the almost vertical right brachiocephalic vein to form the SVC.

11.3 **C.** The anterior mediastinum lies immediately posterior to the sternum and contains the thymus in children.

ANATOMY PEARLS

▶ The surface landmark for the boundary between the superior and inferior (anterior, middle, and posterior) mediastinum is the sternal angle.

▶ The thoracic duct is found in the posterior and superior mediastina.

▶ The right tracheobronchial nodes drain lymph from the right lung and inferior portion of the left lung.

▶ Much of the lymph from the thorax and its contents will drain into the bronchomediastinal trunk.

REFERENCES

Gilroy AM, MacPherson BR, Ross LM. *Atlas of Anatomy*, 2nd ed. New York, NY: Thieme Medical Publishers; 2012:78–79, 85, 89, 114, 127.

Moore KL, Dalley AF, Agur AMR. *Clinically Oriented Anatomy*, 7th ed. Baltimore, MD: Lippincott Williams & Wilkins; 2014:117–118, 127–128, 133, 160–166.

Netter FL. *Atlas of Human Anatomy*, 6th ed. Philadelphia, PA: Saunders; 2014: plates 203, 205.

A 60-year-old woman is noted to have a 2-cm mass in the left breast. The patient's physician recommends that a core needle biopsy be performed. Tissue analysis by the pathologist under the microscope reveals intraductal carcinoma. The patient is advised by the surgeon to have surgery to remove the primary breast mass in addition to some lymph nodes. The patient undergoes wide excision of the breast mass and lymph node removal.

► Which lymph nodes are most likely to be affected?
► What anatomical structure defines the "levels" of lymph nodes?

ANSWER TO CASE 12:

Breast Cancer

Summary: A 60-year-old woman undergoes lumpectomy and lymph node dissection for a 2-cm intraductal carcinoma of the breast.

- **Most likely lymph nodes affected:** Axillary nodes.

- **Anatomical structure that defines the "levels" of lymph nodes:** The pectoralis minor muscle is used to define lymph node levels. Levels 1, 2, and 3 are lateral to, deep to, and medial to the pectoralis minor, respectively.

CLINICAL CORRELATION

This 60-year-old woman had a palpable breast mass. Pathological examination revealed intraductal carcinoma in the core needle biopsy. Risk factors include the patient's age, and intraductal carcinoma is the most common histological type. The most common treatment plan would be a breast-conserving procedure such as a lumpectomy (excising the malignant mass with some margins) and axillary lymph node dissection. The presence or absence of malignant cells in the axillary lymph nodes is the single most important prognostic factor for survival. **Options for nodal staging include level 1 and 2 axillary node dissection versus sentinel node biopsy.** The sentinel node(s) represents the node(s) to which primary lymph drainage occurs from a tumor or anatomical site. It is identified by injection of radiotracers and a blue dye at the primary tumor site. Biopsy of the sentinel node(s) results in a smaller incision and decreased trauma to the axilla. However, if the sentinel node(s) is positive for metastatic disease, a complete level 1 and 2 axillary dissection should be performed.

Other physical signs of breast cancer, which this patient did not have, include skin dimpling or retraction, which is formed by the underlying cancer adherent to the fibrous septa of the breast, and the thickened red appearance of **peau d'orange**, which is caused by the malignant cells proliferating within the lymphatics underlying the skin. A red, warm breast in a non-breast-feeding woman can also represent inflammatory breast cancer due to malignancy within the lymphatic channels of the skin.

APPROACH TO:

The Axillary Lymph Nodes

OBJECTIVES

1. Be able to describe the anatomy of the adult female breast, including the blood and nerve supplies.

2. Be able to list the primary path for lymphatic drainage of the breast and the several subgroups of axillary nodes.

3. Be able to describe the secondary pathways for lymph drainage.

DEFINITIONS

AXILLA: Small pyramidal space between the upper lateral chest and the medial arm, including the blood vessels, nerves, and lymph nodes.

TAIL OF SPENCE: A protrusion of mammary tissue into the axilla that sometimes enlarges premenstrually.

AXILLARY LYMPH NODE DISSECTION: Surgical excision of lymph nodes of the axilla, usually related to breast cancer for diagnostic and therapeutic reasons.

PEAU D'ORANGE: Orange peel appearance of the skin of the breast with edema and prominent pores secondary to obstruction of lymphatics by a tumor with associated inflammation.

DISCUSSION

The adult female breast consists of subcutaneous, radially arranged, mammary gland tissue and fat, typically extending from ribs 2 through 6 superiorly to inferiorly and from the sternal border to the midaxillary line (Figure 12-1). The **long thoracic**

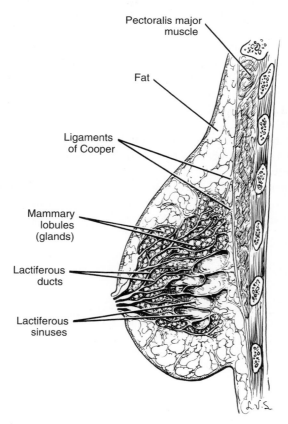

Figure 12-1. Sagittal section of the breast. (*Reproduced, with permission, from Lindner HH. Clinical Anatomy. East Norwalk, CT: Appleton & Lange, 1989:202.*)

nerve lies close to the midaxillary line. For descriptive purposes, it is divided into quadrants. Each breast is centered by the elevated nipple, which contains the openings of the **lactiferous ducts** and is composed of circular smooth muscle. Surrounding the nipple is pigmented skin or the areola, which contains the opening of the lubricating sebaceous glands. The radially arranged mammary gland tissue forms **15 to 20 lobes**, each drained by a lactiferous duct that has a dilatation called the **lactiferous sinus** just before its opening onto the nipple. The lobes are irregularly separated by incomplete dense connective tissue septae that attach to the dermis of the overlying skin. These **septae, called the suspensory ligaments (of Cooper)**, are especially well developed in the superior half of the breast. A loose connective tissue layer, the retromammary space, separates the breast components and the **pectoral fascia**, allowing for some movement. The breast overlies the pectoralis major and the anterior portion of the serratus anterior muscles. A portion of breast tissue typically extends into the axilla as the **axillary tail (of Spence)**. The breast is supplied by branches of the **internal thoracic, lateral thoracic, and anterior and posterior intercostal arteries. The breast is innervated by anterior and lateral cutaneous branches of intercostal nerves.**

Lymphatic drainage of the breast begins as a subareolar plexus. The majority of lymph drained from the breast (usually quantified at 75 percent) **drains to the axillary lymph nodes.** The axillary node group is often described as a pyramid, like the axilla, and is typically subdivided into **five subgroups: pectoral (anterior), lateral (humeral), posterior (subscapular), central (medial), and apical.** Lymph from the axillary nodes typically drains into the **inferior deep cervical lymph nodes.** However, lymph from the axillary node group may drain into other nodes such as the interpectoral and deltopectoral nodes (Figure 12-2). This is especially true in instances of metastasis because "normal" paths become blocked by the malignancy and alternate routes must be established. **The pectoral, humeral, and subscapular nodes are level 1 nodes, whereas the central and apical nodes are level 2 and 3 nodes, respectively.**

The **medial quadrants of the breast will have lymph drain into the parasternal lymph nodes** along the **internal thoracic vessels.** Some lymph from the inferior quadrants may drain to **inferior phrenic nodes.**

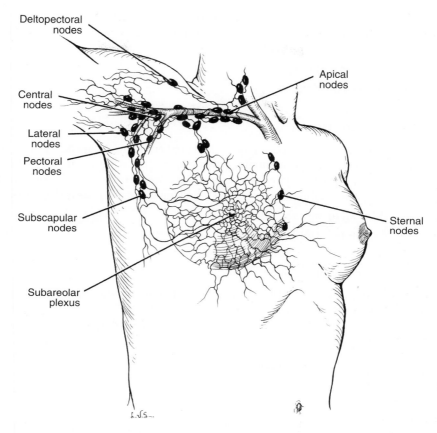

Figure 12-2. Lymphatics of the breast. (*Reproduced, with permission, from Lindner HH. Clinical Anatomy. East Norwalk, CT: Appleton & Lange, 1989:205.*)

COMPREHENSION QUESTIONS

12.1 A 45-year-old woman is noted to have a 1.5-cm breast cancer located in the upper inner quadrant of the right breast. Which of the following lymph nodes is most likely to be affected?

 A. Level 1 axillary node

 B. Level 2 axillary node

 C. Level 3 axially node

 D. Parasternal node

 E. Inferior phrenic node

12.2 A physician is performing a breast examination. In addition to the breast tissue on the chest, what other region is critical to complete the palpation of mammary tissue?

A. Supraclavicular region
B. Subclavicular region
C. Axillary region
D. Parasternal region

12.3 A 24-year-old woman has vaginally delivered an infant 2 days ago. She complains of breast engorgement and swelling in regions at about the level of the umbilicus and at the lateral abdomen. There seems to be some leaking from these areas of swelling. Which of the following is the most likely diagnosis?

A. Bilateral lipoma
B. Accessory breast tissue
C. Ascites
D. Cutaneous malignancy

12.4 A 52-year-old woman undergoes surgery and finds that she cannot abduct her left arm past 90 degrees. Also, on examination, her left scapula is abnormally prominent. Which surgery is most likely responsible for the patient's condition?

A. Left lumpectomy and sentinel node biopsy
B. Left radical mastectomy
C. Left carotid artery endarterectomy
D. Splenectomy

ANSWERS

12.1 **D.** Cancers located in the medial breast usually drain to the parasternal nodes.

12.2 **C.** The tail of Spence is located in the axillary area and contains mammary tissue.

12.3 **B.** These areas likely are accessory breast tissue. The "milk line" extends from the axilla to the groin area, and accessory mammary tissue may be present anywhere along this line.

12.4 **B.** The patient likely has an injury to the left long thoracic nerve, and the deficits resulting from weakness to the left serratus anterior muscle. Injury to the long thoracic nerve leads to inability to abduct the arm past 90 degrees, and also the appearance of a "winged scapula." A radical mastectomy is the most likely surgery to lead to injury to the long thoracic nerve, particularly in the axillary region. A sentinel node biopsy is a less extensive surgery and not as likely to injure this nerve.

ANATOMY PEARLS

▶ The breast is a subcutaneous structure composed of 15 to 20 lobes of mammary gland tissue and fat and typically extends into the axilla as the axillary tail.

▶ The breast extends from ribs 2 through 6 and from the sternal border to the midaxillary line. This places the lateral thoracic nerve at risk during surgery.

▶ The suspensory ligaments of the breast are attached to the dermis of the skin.

▶ The majority of lymph from the breast drains to the axillary lymph nodes, with secondary drainage to the parasternal and inferior phrenic nodes.

REFERENCES

Gilroy AM, MacPherson BR, Ross LM. *Atlas of Anatomy*, 2nd ed. New York, NY: Thieme Medical Publishers; 2012:72–73.

Moore KL, Dalley AF, Agur AMR. *Clinically Oriented Anatomy*, 7th ed. Baltimore, MD: Lippincott Williams & Wilkins; 2014:98–101, 104–106.

Netter FH. *Atlas of Human Anatomy*, 6th ed. Philadelphia, PA: Saunders; 2014: plates 179–182.

CASE 13

A 35-year-old Hispanic woman comes to your office tired and complaining of shortness of breath and fatigue. Her history is unremarkable except for a vague history of fever and joint pain as a child in Mexico. She notes some recent fatigue and difficulty sleeping that she attributes to job-related stress. On examination, her heart rate is 120 beats/min, and the rhythm has no discernible pattern (is irregularly irregular). Auscultation of the heart indicates a systolic murmur (during left ventricular ejection of blood) that is harsh in character.

▶ What is the most likely diagnosis?
▶ What is the underlying etiology?

ANSWER TO CASE 13:
Atrial Fibrillation/Mitral Stenosis

Summary: A 35-year-old Hispanic woman complains of fatigue. She had fever and joint pain as a child in Mexico. On examination, her heart rate is 120 beats/min and irregularly irregular. Cardiac examination shows a harsh systolic murmur.

- **Most likely diagnosis:** Atrial fibrillation due to left atrial enlargement

- **Underlying etiology:** Mitral stenosis due to rheumatic heart disease

CLINICAL CORRELATION

This 35-year-old woman most likely has atrial fibrillation with tachycardia that is irregularly irregular. The electrical impulse originating from the sinoatrial (SA) node of the right atrium does not depolarize both atria in a regular, orderly manner; instead, this patient's atria receive constant electrical stimulation, leading to almost continual atrial contraction that visually resembles a bag of worms. The irregular character of the pulse is the result of inconsistent transmission of the electrical impulse to and through the atrioventricular (AV) node and then onto the two ventricles. One common cause of atrial fibrillation is left atrial enlargement. In this patient, the history of childhood fever and joint pain likely is the result of streptococcally caused rheumatic fever. If untreated, the microorganism can cause inflammation of the mitral valve, leading to mitral stenosis. After 3-5 years, the mitral stenosis is likely to worsen, leading to atrial enlargement, fibrillation, and pulmonary edema with intolerance to physical exertion. Treatment in this patient would focus on decreasing her heart rate with an agent that acts on the AV node such as digoxin. Oxygen and diuretics would relieve her pulmonary symptoms. An ultimate goal will be conversion of her cardiac contractions to a normal sinus rhythm. Anticoagulation is often warranted in the face of long-term atrial fibrillation because of the likelihood of intracardiac thrombus and the possibility of emboli after conversion to sinus rhythm, called the "atrial stunning" effect. Surgical correction of the mitral stenosis is also important.

APPROACH TO:
Cardiac Conduction System

OBJECTIVES

1. Be able to describe the type of tissue that makes up the cardiac conduction system

2. Be able to describe the locations and functions of the SA node, the AV node, the AV bundle (of His), and the right and left bundle branches

3. Be able to describe the nature of sinus rhythm and the influence of the divisions of the autonomic nervous system on this rhythm

4. Be able to describe the anatomy of the four cardiac valves

DEFINITIONS

MURMURS: Soft or harsh abnormal heart sounds, often caused by turbulent blood flow, and described in relation to the phase of the cardiac cycle in which they are heard

ATRIAL FIBRILLATION: Rapid, uncoordinated muscular twitching of the atrial wall

TACHYCARDIA: A heart rate of at least 100 beats/min

DISCUSSION

Cardiac Conduction System

The **conduction system of the heart** is composed of specially modified cardiac muscle cells. It initiates and rapidly conducts cardiac impulses throughout the heart to produce cardiac muscle contraction. The system ensures the simultaneous contraction of both atria, followed by a similar coordinated contraction of both ventricles.

The **SA node,** composed of these modified cardiac muscle cells, lies within the **atrial wall on the right side** of its junction with the **superior vena cava** (SVC). This can be located at the superior end of the external landmark, the **sulcus terminalis.** The SA node spontaneously depolarizes to initiate the cardiac conduction impulse and thus is often referred to as the **heart's pacemaker.** The impulse generated by the SA node spreads through the atrial wall to converge on the **AV node** and produces simultaneous atrial contraction. **Anterior, middle, and posterior internodal pathways** of very rapid conduction are described (Figure 13-1).

The **AV node** is a somewhat smaller mass of modified cardiac muscle cells located in the **interatrial septum,** immediately superior to the opening of the **coronary sinus.** The **AV bundle (of His)** arises from this node and lies within the **membranous portion of the interventricular septum.** It courses toward the apex of the heart, and at the upper portion of the muscular portion of this septum, it divides into **right and left bundle branches.** The bundle branches lie on their respective sides of the septum just beneath the endocardium. The bundles then divide to form a **subendocardial plexus of Purkinje fibers.** The right bundle is described as supplying the interventricular septum, the anterior papillary muscle (reached by the septomarginal or moderator band), and the wall of the right ventricle. The left bundle supplies the interventricular septum, anterior and posterior papillary muscles, and the wall of the left ventricle.

The SA node or pacemaker typically will depolarize at a rate of approximately 70 times per minute. This rate is referred to as a **sinus rhythm.** The SA node is innervated by fibers of the **sympathetic and parasympathetic divisions** of the autonomic nervous system. Stimulation of the SA node by sympathetic nerve impulses increases the rate of depolarization of the SA node, and stimulation by parasympathetic fibers decreases this rate.

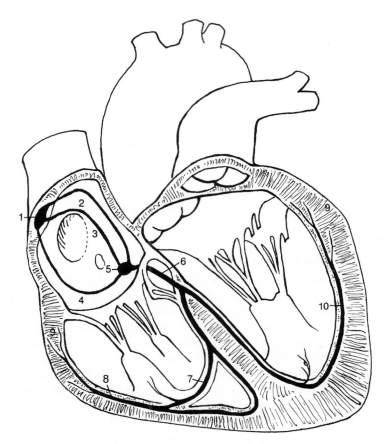

Figure 13-1. Cardiac conduction system: 1 = sinoatrial node, 2 = anterior internodal pathway, 3 = middle internodal pathway (Wenckebach bundle), 4 = posterior internodal pathway, 5 = atrioventricular node, 6 = atrioventricular bundle of His, 7 = moderator band, 8 = right bundle branch, 9 = terminal conducting fibers of Purkinje, 10 = left bundle branch. (*Reproduced, with permission, from the University of Texas Health Science Center Houston Medical School.*)

Cardiac Valves

The outflow from the two atria and the two ventricles is guarded by the **AV and the semilunar valves,** respectively. The leaflets of these cardiac valves and the myocardial muscle fibers are attached to the **fibrous cardiac skeleton.** This structure consists of **four fibrous rings** to which the leaflets attach, the right and left fibrous trigone, and the membranous portion of the interventricular septum.

The right AV or tricuspid valve between the right atrium and right ventricle consists of anterior, posterior, and septal leaflets or cusps. **Tendinous cords** attach to the margins of adjacent valve cusps and prevent separation and inversion (prolapse) of the leaflets into the atrium during ventricular contraction. The proximal attachment of the tendinous cords is to conical projections of cardiac muscle called **papillary muscles;** there are three papillary muscles, named **anterior, posterior,** and **septal,** like the cusps. The tendinous cords of the anterior papillary muscle attach to the anterior and posterior cusps. Those of the posterior papillary muscle attach to the posterior and septal cusps, and the cords of the septal papillary muscle attach to the septal and anterior cusps.

The left AV or **bicuspid (mitral) valve** between the left atrium and ventricle consists of anterior and posterior cusps. The tendinous cords of the anterior and posterior papillary muscles (which are larger because of the increased pressure demands) are attached to adjacent cusps and function in a manner similar to that described for the tricuspid valve.

The outflow from the right and left ventricles is guarded by the **pulmonary and aortic semilunar valves,** respectively. Both semilunar valves are similar in structure; both are circular in shape and consist of three cuplike cusps, with the opening to these cups directed superiorly. The space formed is called the **pulmonary or aortic sinus,** so named for the cusp that creates it. As blood is ejected from the ventricles, the cusps lie close to the pulmonary or aortic wall. At the end of contraction, the elasticity of the vessel walls results in backflow of blood that fills the sinuses, resulting in apposition of the three cusps and closure of the valves. The cusps of the pulmonary semilunar valve are the anterior, right, and left cusps, and the aortic valve has right, left, and posterior cusps. The right and left coronary arteries arise from the aorta at the right and left aortic sinuses, respectively.

COMPREHENSION QUESTIONS

13.1   As a pathologist, you are examining the heart of a victim of fatal trauma and note a tear at the junction of the SVC and the right atrium. This tear would likely damage which of the following?

    A.   SA node

    B.   AV node

    C.   AV bundle

    D.   Right bundle branch

    E.   Left bundle branch

13.2   As a pathologist, you must examine the AV bundle histologically. In which of the following tissue samples will you find the AV bundle?

    A.   Right atrium

    B.   Left atrium

    C.   Interatrial septum

    D.   Membranous interventricular septum

    E.   Muscular interventricular septum

13.3 A 57-year-old man develops a myocardial infarction and is noted to have a heart rate of 40 beats/min. The cardiologist diagnoses an occlusion of the right coronary artery. Which of the following structures is most likely to be affected?

A. AV node

B. Bundle of His

C. Purkinje fibers

D. Mitral valve

ANSWERS

13.1 **A.** The SA node or pacemaker lies within the right atrial wall, where it is joined by the SVC.

13.2 **D.** The AV bundle is located in the membranous portion of the interventricular septum.

13.3 **A.** An inferior wall myocardial infarction involving the right coronary artery may affect the AV node, leading to bradycardia.

ANATOMY PEARLS

▶ The cardiac conduction system is composed of specially modified cardiac muscle cells (not nervous tissue).

▶ The SA node is the pacemaker that spontaneously produces a sinus rhythm of 70 beats/min. It lies at the junction of the SVC and the right atrium.

▶ The AV node lies in the interatrial septum, and the AV bundle and the right and left bundle branches lie in the membranous and muscular portions of the interventricular septum, respectively.

▶ Stimulation of the SA node by sympathetic nerve impulses increases its rate of depolarization, whereas parasympathetic impulses decrease its depolarization rate.

REFERENCES

Gilroy AM, MacPherson BR, Ross LM. *Atlas of Anatomy*, 2nd ed. New York, NY: Thieme Medical Publishers; 2012:89–93.

Moore KL, Dalley AF, Agur AMR. *Clinically Oriented Anatomy*, 7th ed. Baltimore, MD: Lippincott Williams & Wilkins; 2014:135–150, 159.

Netter FH. *Atlas of Human Anatomy*, 6th ed. Philadelphia, PA: Saunders; 2014: plates 215–223.

A 65-year-old woman diagnosed with uterine cancer underwent surgery to remove her uterus (total abdominal hysterectomy) 2 days previously. She was doing well until today, when she developed shortness of breath, and she describes a sharp pain in the right side of her chest on inspiration. Physical examination revealed a respiratory rate of 28 breaths/min and a heart rate of 110 beats/min. Auscultation of the lungs demonstrates no wheezing or crackles. She appears anxious.

▶ What is the most likely diagnosis?
▶ What is the most likely location of the primary disease?

ANSWER TO CASE 14:
Pulmonary Embolism

Summary: Two days ago, a 65-year-old woman underwent a total abdominal hysterectomy because of endometrial cancer. She developed acute-onset dyspnea with pleuritic chest pain. She has tachypnea and tachycardia and appears anxious. Rales (crackles) are not present on pulmonary examination.

- **Most likely diagnosis:** Pulmonary embolism

- **Most likely location of the primary disease:** Deep-vein thrombosis (DVT) of the pelvis or lower limb

CLINICAL CORRELATION

This woman has multiple risk factors for DVT or blood clot formation within the large veins. These factors include the patient's age, likely minimal physical exercise, and bedrest after a major operative gynecological procedure for a cancerous lesion. Postoperative orthopedic patients are similarly at risk. Deep-vein thrombi are typically asymptomatic but may cause lower-limb swelling and pain. When pelvic or lower limb veins are involved, clot material can break free (embolize) and travel through the **inferior vena cava** (IVC) to and through the right side of the heart, whence they are pumped to the lungs, where they will lodge in branches of the pulmonary arteries. These emboli effectively block blood flow beyond this point and prevent this unoxygenated blood from reaching the alveoli, where it is to be oxygenated. The size and number of emboli produced will determine the amount of lung tissue that will be infarcted because of lack of oxygen. The most common symptom of pulmonary embolism is dyspnea, and patients are often anxious, with tachycardia and pleuritic chest pain at inspiration. The next step would be an arterial blood-gas study to assess oxygen status. A chest radiograph and ventilation-perfusion scan are performed to directly determine whether an embolus is present. If present, intravenous anticoagulants such as heparin are beneficial. Large or untreated emboli can cause death. One particularly devastating type, the "saddle embolus," lodges in the pulmonary trunk at the bifurcation of the right and left pulmonary arteries, thus blocking blood flow to both lungs, leading to cardiovascular collapse and death.

APPROACH TO:
Pulmonary Vasculature

OBJECTIVES

1. Be able to describe the origin, branching pattern, and anatomical relations of the pulmonary arteries and veins

2. Be able to describe the origin of the bronchial arteries, structures supplied, and sites of anastomosis with the pulmonary circulation

DEFINITIONS

TOTAL HYSTERECTOMY: Complete surgical removal of the uterus, that is, the body and the cervix. A subtotal hysterectomy consists in removal of the uterine corpus (body) but not of the cervix.

PULMONARY EMBOLISM: Obstruction or occlusion of pulmonary arteries by emboli typically arising from thrombi of veins in the lower limbs or the pelvis.

INFARCTION: Tissue necrosis due to the sudden decrease in the blood supply as the result of an embolus, thrombus, or external pressure.

RALES: "Crackles" heard when listening to the lung fields with a stethoscope, usually indicative of excess fluid in the lungs as with pneumonia or pulmonary edema.

DISCUSSION

The pulmonary trunk, which carries unoxygenated blood, arises from the **conus arteriosus** portion of the right ventricle. At the level of the sternal angle, the trunk divides into **right and left pulmonary arteries** (see Figure 14-1). The right pulmonary artery passes laterally, posterior to the ascending aorta and SVC, to reach the hilum of the right lung. The left pulmonary artery passes anterior to the descending thoracic aorta to reach the hilum of the left lung. The pulmonary arteries are the most superior vessels in the hilum of each lung, and the branch to the superior lobe of each lung typically arises outside the lung hilum. Each artery courses through the lung tissue adjacent to bronchial and bronchiolar airway structures, where they

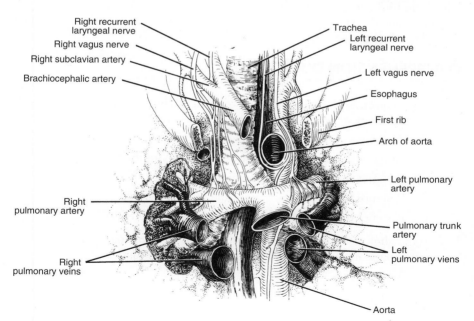

Figure 14-1. Superior mediastinum and relations of the pulmonary vessels. (*Reproduced, with permission, from Way LW, ed. Current Surgical Diagnosis and Treatment, 7th ed. East Norwalk, CT: Appleton & Lange, 1985.*)

branch out and are named for these airway structures. Thus each artery will divide into lobar and then segmental branches to the lung lobes and their **bronchopulmonary segments,** respectively. The **bronchioles** and the adjacent arteries branch further down to the level of the **terminal bronchiole,** which supplies a lobule, the smallest anatomical unit of lung tissue.

As the small pulmonary artery branches reach the respiratory bronchioles, they form the extensive capillary network around and between the alveoli. The **thin capillary endothelium, basal lamina, and type I pneumocytes form the blood-gas barrier through which gaseous exchange occurs.**

Oxygenated blood drains from the capillary bed to pulmonary veins within the thin connective tissue septae between lobules. In this location, they receive blood from adjacent lobules. As the pulmonary veins unite to form increasingly larger veins, they remain separated from the pulmonary artery and airway structures; they are found at the periphery of lung tissue subdivisions such as the bronchopulmonary segments and lobes. These larger veins will drain adjacent segments or lobes. Eventually, two pulmonary veins exit the hilum of each lung anteriorly and inferiorly to the entering pulmonary arteries. Thus **four pulmonary veins** drain oxygenated blood into the left atrium, typically two from each lung.

The **bronchi, bronchioles,** and related structures, the connective tissue stroma and the visceral pleura, receive their blood supply from bronchial arteries. These are typically branches of the **thoracic aorta** but may arise from intercostal arteries. Anastomoses between the pulmonary and bronchial arteries occur within the bronchial walls and the visceral pleura. Bronchial veins from the right and left lungs typically drain to the azygous and accessory hemiazygous veins, respectively, but carry only small amounts of blood. The pulmonary vein carries most of the blood supplied by the bronchial arteries.

COMPREHENSION QUESTIONS

14.1 As a surgeon exploring the thorax, you will be able to identify the right pulmonary artery in which of the following locations?

  A. Anterior to the ascending aorta and the SVC

  B. Anterior to the ascending aorta and posterior to the SVC

  C. Posterior to the descending aorta and the SVC

  D. Posterior to the ascending aorta and the SVC

  E. Posterior to the ascending aorta and anterior to the SVC

14.2 As a radiologist examining a contrast study of the pulmonary vessels, you will note how many pulmonary veins entering the left atrium?

  A. Two

  B. Three

  C. Four

  D. Five

  E. Six

14.3 A 44-year-old woman who has a DVT of the lower extremity suddenly gasps and collapses. She is found to be hypotensive. Resuscitative measures are attempted without success. Which of the following is the most likely diagnosis?

A. Myocardial infarction
B. Saddle embolus
C. Right peripheral pulmonary embolus
D. Embolic stroke

ANSWERS

14.1 **D.** The right pulmonary passes posteriorly to the ascending aorta and the SVC.

14.2 **C.** Four pulmonary veins that carry oxygenated blood drain into the left atrium.

14.3 **B.** The patient likely developed a saddle embolus that occluded blood flow to both pulmonary arteries and, in effect, stopped the circulatory system.

ANATOMY PEARLS

▶ Pulmonary arteries carry unoxygenated blood, accompany airway structures, and follow their branching patterns

▶ Pulmonary veins, which carry oxygenated blood, course separately from the arteries and airways at the periphery of lung tissue subdivisions.

▶ The blood-gas barrier is composed of the capillary endothelium, basal lamina, and type I pneumocytes.

▶ Bronchial arteries typically arise from the thoracic aorta and supply the airway structures and stromal tissue.

REFERENCES

Gilmore AM, MacPherson BR, Ross LM. *Atlas of Anatomy*, 2nd ed New York, NY: Thieme Medical Publishers; 2012:88, 124–126.

Moore KL, Dalley AF, Agur AMR. *Clinically Oriented Anatomy*, 7th ed. Baltimore, MD: Lippincott Williams & Wilkins; 2014:116–117, 124–125.

Netter FH. *Atlas of Human Anatomy*, 6th ed. Philadelphia, PA: Saunders; 2014: plates 202–204.

A 54-year-old man who had smoked two packs of cigarettes per day for 20 years complains of the acute onset of shortness of breath, and severe chest pain with respiratory movement. Physical examination reveals a barrel chest consistent with chronic obstructive pulmonary disease. There are decreased breath sounds on the right side. When the physician taps on the right chest (percussion), there is a hyperresonant (unusually hollow) sound.

► What is the most likely diagnosis?
► What is the anatomical disorder?

ANSWER TO CASE 15:

Pneumothorax

Summary: A 54-year-old smoker complains of acute-onset shortness of breath and severe chest pain with breathing. He has physical findings for chronic obstructive pulmonary disease with a barrel chest. There are decreased breath sounds and hyper-resonance to percussion on the right side.

- **Most likely diagnosis:** Pneumothorax

- **Anatomical disorder:** Entry of air into the pleural space, resulting in lung collapse

CLINICAL CORRELATION

Air is drawn into the lungs through the trachea and bronchi by the increasing negative thoracic pressure produced by the downward movement of the diaphragm. If air enters the pleural space through the thoracic wall or the surface of the lung itself, the negative pressure of the pleural space equilibrates with atmospheric pressure, and air movement ceases. The defect that allowed air to enter the pleural space acts like a valve by preventing the air from exiting the space. Pressure increases above that of atmospheric pressure, and a **tension pneumothorax** results, which is characterized by lung collapse, with displacement toward the mediastinum. A severe pneumothorax may cause displacement of the mediastinum and its contents toward the intact lung and partial compression of this lung. The most serious consequence of these anatomical shifts is decreased venous return to the heart. A patient who has chronic obstructive pulmonary disease is at risk for spontaneous pneumothorax by the rupture of an emphysematous bleb on the surface of the lung. Spontaneous pneumothorax may also occur from lung surface blebs in young men. The typical clinical presentation of pneumothorax is chest pain with dyspnea, decreased breath sounds, and hyperresonance on the affected side. The diagnosis is confirmed by chest radiograph. Treatment is directed toward removal of the air from the pleural space with a needle in emergent situations or by a chest tube placed in the pleural space and directed to an underwater seal.

APPROACH TO:

The Pleural Cavities

OBJECTIVES

1. Be able to describe the contents of the pulmonary cavities: lungs and the pleural divisions

2. Be able to describe the superior and inferior limits of the pleural cavity and the lower limits of each lung

3. Be able to describe the functional importance of the pleural cavity and fluid and the pressure within the cavity

DEFINITIONS

CHRONIC OBSTRUCTIVE PULMONARY DISEASE: General term applied to permanent or temporary diseases that cause narrowing of the bronchi so as to obstruct forced expiratory flow; includes bronchitis, emphysema, and asthma

EMPHYSEMA: A lung condition in which the air spaces distal to the terminal bronchioles are larger than normal

PNEUMOTHORAX: Air or gas within the pleural cavity

CHEST TUBE: Tube inserted through the thoracic wall into the pleural cavity for the purpose of draining air or fluid from that cavity

DISCUSSION

The **skeletal components of the thoracic wall** are the thoracic vertebra, the attached **12 pairs of ribs, and the sternum.** The interval between the ribs is closed by three layers of muscles: the **external, internal, and innermost intercostal muscles.** The innermost intercostal is largely laterally located, internal to the internal intercostal. The transverse thoracic and subcostal muscles are discontinuous thoracic wall muscles found anteriorly and laterally, respectively. Externally, several muscles associated with the upper limb or accessory respiratory muscles attach to the thoracic wall. These include the pectoralis major and minor, serratus anterior and posterior, scalene, and levator costarum muscles. Internally, the thoracic cavity is divided into two pulmonary cavities separated by the central **mediastinum.** The thoracic cavity is closed inferiorly by the diaphragm.

Each of the two laterally placed pulmonary cavities contains a lung covered with visceral pleura, and each is lined with parietal pleura. The two pleura are continuous with each other at the root of the lung, where neurovascular and airway structures enter and exit the lung. The **pleura** is a serous membrane composed of mesothelium and a small amount of connective tissue, and it produces the lubricating pleural fluid. The parietal pleura is divided for descriptive purposes into four parts according to the structure to which it is attached. The costal, diaphragmatic, mediastinal, and cervical portions are attached, respectively, to the inner aspect of the thoracic wall; the superior surface of the diaphragm; the lateral aspect of the mediastinum, especially the pericardial sac; and the root of the neck superior to the superior thoracic aperture.

Relative constant pleural lines of reflection are created as one portion of parietal pleura changes direction to attach to another structure (Figure 15-1). The **sternal reflection line** is created as the mediastinal pleura changes direction (is reflected) onto the inner thoracic wall and becomes the costal pleura. In the right pulmonary cavity, this line of reflection is close to the midline from the sternal angle to the xiphoid process. On the left side, the line of reflection courses from the sternal angle to the level of the fourth rib, and then arches to the left to the sixth rib in the midclavicular line, thus creating the cardiac notch. The curvature of the mediastinal surface in this region results in the formation of a shallow costomediastinal recess of the pleural cavity. Inferiorly, as the **costal pleura** are reflected onto the surface of the diaphragm, the costal reflection line is created. The surface landmarks for

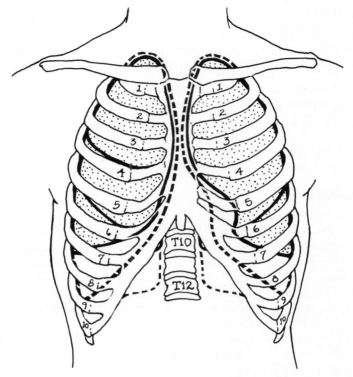

Figure 15-1. The lungs (bounded by solid line) and pleura (denoted by the heavy dotted line). (*Reproduced, with permission, from the University of Texas Health Science Center Houston Medical School.*)

this reflection line on the right and left sides are the 8th rib at the midclavicular line (MCL), the 10th rib at the midaxillary line (MAL), and the 12th rib at the vertebral border. These landmarks also mark the inferior limits of the pleural cavity. The curved shape of the diaphragmatic pleura on the dome of the diaphragm and the vertical costal pleura form a wedge-shaped pleural cavity recess called the **costo-diaphragmatic recess,** in which abnormal pleural cavity fluids such as blood or pus will accumulate. The lowest level of each lung at the end of expiration is the 6th rib at the MCL, the 8th rib at the MAL, and the 10th rib at the vertebral border. The cervical pleura and thus the pleural cavity extend into the root of the neck, 2 to 3 cm superior to the medial end of the clavicle.

The **pleural cavity between the visceral and parietal layers of pleura** is a potential space containing a small amount of lubricating pleural fluid. This fluid wets the surface of the lungs, resulting in adherence of the lung's visceral pleura to the costal and diaphragmatic parietal pleura by surface tension forces. As the diaphragm descends and the thoracic wall expands with inspiration, the adherent lungs also expand. The pleural cavities are completely closed spaces and are at 756 mmHg of pressure, or at −4 mmHg with respect to atmospheric pressure (760 mmHg). If the visceral pleura covering the lung is ruptured or the costal parietal pleura is disrupted by trauma, air will enter the pleural cavity, causing a pneumothorax and at least equalizing pleural pressure with atmospheric pressure. This will produce at least a partial lung collapse and interfere with ventilation and gaseous exchange.

COMPREHENSION QUESTIONS

15.1 You must remove fluid from the pleural cavity of your patient (thoracentesis). You decide to insert the aspiration needle over the top of a rib, into an intercostal space inferior to the lower border of the lung in the MAL at the end of a normal expiration. Which of the following is the lowest (most caudal) level at which this procedure might safely be done without injuring the lung?

A. Fourth intercostal space

B. Fifth intercostal space

C. Sixth intercostal space

D. Seventh intercostal space

E. Eighth intercostal space

15.2 During this thoracentesis procedure, the lowest level of the pleural cavity will lie at the level of which rib at the end of expiration in the MAL?

A. Seventh

B. Eighth

C. Ninth

D. Tenth

E. Eleventh

15.3 During this procedure, the lower border of the lung will lie at the level of which rib in the MCL?

A. Fifth

B. Sixth

C. Seventh

D. Eighth

E. Ninth

ANSWERS

15.1 **E.** The lower border of the lung will lie at the level of the eighth rib in the MAL, enabling safe insertion of the needle into the eighth intercostal space.

15.2 **D.** The lowest level of the pleural cavity in the MAL lies at the level of the 10th rib.

15.3 **B.** The lower border of the lung at the MCL at the level of the sixth rib.

ANATOMY PEARLS

► Visceral and parietal pleura are continuous with each other at the root of the lung.

► The inferior extent of the pleural cavity is the 8th rib at the MCL line, the 10th rib at the MAL, and 12th rib at the vertebral border.

► The inferior border of each lung at the end of expiration is the 6th rib at the midclavicular line, the 8th rib at the MAL, and the 10th rib at the vertebral border.

► The pleural cavity is at -4 mmHg with respect to atmospheric pressure.

REFERENCES

Gilroy AM, MacPherson BR, Ross LM. *Atlas of Anatomy*, 2nd ed. New York, NY: Thieme Medical Publishers; 2012:110–114.

Moore KL, Dalley AF, Agur AMR. *Clinically Oriented Anatomy*, 7th ed. Baltimore, MD: Lippincott Williams & Wilkins; 2014:106–113, 121.

Netter FH. *Atlas of Human Anatomy*, 6th ed. Philadelphia, PA: Saunders; 2014: plates 193–195.

A 59-year-old man complains of tight chest pressure and shortness of breath after lifting several boxes in his garage approximately 2 h ago. He perceives that his heart is skipping beats. His medical history is significant for hypertension and cigarette smoking. On examination, his heart rate is 55 beats/min and regular, and his lungs are clear to auscultation. An electrocardiogram shows bradycardia with an increased PR interval and ST-segment elevation in multiple leads including the anterior leads, V1 and V2.

▶ What is the most likely diagnosis?
▶ What anatomical structures are most likely affected?

ANSWER TO CASE 16:

Coronary Artery Disease

Summary: A 59-year-old hypertensive male smoker has a 2-h history of tight chest pressure, shortness of breath, and palpitations after exertion. His heart rate is 55 beats/min and regular. The electrocardiogram (ECG) shows bradycardia, first-degree heart block, and ST-segment elevation in leads V1 and V2.

- **Most likely diagnosis:** Myocardial infarction

- **Anatomical structures likely affected:** Right coronary artery and left anterior descending artery

CLINICAL CORRELATION

This patient's 2-h history of worsening chest pain, dyspnea, and palpitations after physical exertion is classic for a myocardial infarction. The pain of angina due to the myocardial ischemia is typically deep, visceral, and squeezing in nature, like an "elephant stepping on the chest." It frequently radiates to the neck or left arm. This patient's risk factors include hypertension and tobacco use. The ECG (ST-segment elevation) is highly suspicious for myocardial infarction. Leads V1 and V2 are used to evaluate the anterior portion of the heart, which is supplied by the left anterior descending artery. Bradycardia and first-degree heart block (increased PR interval) indicate right coronary artery disease.

APPROACH TO:

Coronary Artery Circulation

OBJECTIVES

1. Be able to describe the course and areas of the heart supplied by the right and left coronary arteries, respectively

2. Be able to describe the venous drainage of the heart

3. Be able to describe the arterial supply and venous drainage of the pericardial sac

DEFINITIONS

ANGINA: Chest pain classically described as pressure or squeezing indicative of coronary artery insufficiency and cardiac ischemia

ISCHEMIA: Inadequate blood supply and oxygen delivery to tissue

PALPITATIONS: Pulsations of the heart perceptible by a patient that are usually irregular and increased in force

BRADYCARDIA: Heart rate no higher than 60 beats/min

DISCUSSION

The **heart** receives its **arterial blood supply** from the **first branches of the ascending aorta,** the **right and left coronary arteries.** The right and left arteries arise from the aorta at the aortic sinuses, the pockets formed by the right and left cusps of the aortic valve, respectively. Each artery will supply portions of the atria and ventricles.

The **right coronary artery (RCA)** arises at the right aortic sinus and courses in the coronary (AV) groove between the right atrium and ventricle. At the level of the right auricular appendage, it gives off the SA nodal branch, which ascends to the junction of the SVC with the right atrium, where the SA node is located. As it reaches the inferior margin of the heart in the coronary groove, it will usually give off a right marginal branch that supplies the right ventricle along the inferior border. The RCA then curves around the inferior margin of the heart in the coronary groove onto the inferior and posterior surfaces of the heart, passing somewhat to the left toward the junction with the posterior interventricular groove, also called the **crux of the heart.** At the crux, the AV nodal branch passes deep into the interatrial septum to supply the AV node. The RCA divides into a larger posterior interventricular artery, which descends in the groove or sulcus of the same name. It passes toward but typically does not reach the apex of the heart. It supplies the right and left ventricles and posterior portions of the interventricular septum. A small branch continues to the left side of the heart to supply portions of the left atrium and ventricle and will anastomose with the circumflex branch of the left coronary artery (LCA) (see Figures 16-1 and 16-2).

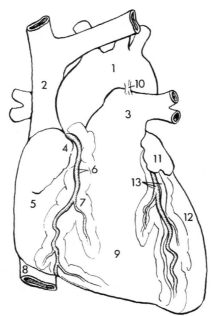

Figure 16-1. Anterior view of the heart: 1 = aortic arch, 2 = superior vena cava, 3 = pulmonary trunk, 4 = right auricle, 5 = right atrium, 6 = core͓   ͓us and vessels, 7 = epicardial fat, 8 = inferior vena cava, 9 = right ventricle, 10 = ligamentum arteriosum, 11 = left auricle, 12 = left ventricle, 13 = anterior interventricular sulcus and vessels. (*Reproduced, with permission, from the University of Texas Health Science Center Houston Medical School.*)

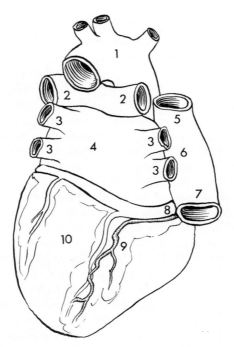

Figure 16-2. Posterior view of the heart: 1 = aortic arch, 2 = pulmonary arteries, 3 = pulmonary veins, 4 = left atrium, 5 = superior vena cava, 6 = right atrium, 7 = inferior vena cava, 8 = coronary sinus, 9 = posterior interventricular sulcus and vessels, 10 = left ventricle. (*Reproduced, with permission, from the University of Texas Health Science Center Houston Medical School.*)

The **LCA** arises from the left aortic sinus and quickly bifurcates into an anterior interventricular and circumflex arteries. The anterior interventricular or left anterior descending (LAD) artery descends toward the apex of the heart in the anterior interventricular groove, where it curves around the apex onto the diaphragmatic surface of the heart to anastomose with the posterior (descending) interventricular branch of the RCA. The anterior interventricular artery supplies the anterior portion of the right and left ventricles and anterior two-thirds of the interventricular septum and therefore is the chief blood supply to the AV and the right and left bundles of the heart's conducting system. The other, smaller, branch of the LCA is the **circumflex branch,** which travels in the coronary groove toward the left margin of the heart, at which point it typically gives off a left marginal branch that supplies the left heart border portion of the left ventricle. The circumflex artery curves around the left heart border to anastomose with the RCA at the posterior aspects of the left atrium and ventricle. The pattern of arterial blood supply at this point is often described as a balanced blood supply because the RCA and LCA supply approximately equal amounts to the heart. In approximately 15 percent of the population, the LCA will supply a larger proportion than the RCA.

The majority of **venous blood will enter the right atrium** through the coronary sinus, which lies in the coronary groove on the posterior surface of the heart. Its internal opening is adjacent to the opening of the IVC. Great, middle, and small cardiac veins and several smaller named veins drain into the coronary sinus. A variable number of small anterior cardiac veins drain directly into the right atrium. The smallest cardiac veins drain small amounts of blood from the myocardial capillary plexus directly into the atria and ventricles.

The heart's **pericardial sac** receives its arterial blood supply primarily from the **pericardiacophrenic artery** (a branch of the internal thoracic artery) that accompanies the phrenic nerve. Small amounts of arterial blood are also provided by branches of the musculophrenic, superior phrenic, bronchial, and esophageal arteries. Pericardiacophrenic veins drain blood to the internal thoracic or brachiocephalic veins.

COMPREHENSION QUESTIONS

16.1 As a cardiologist, you are concerned about blockage of the artery to the SA node in a patient. This artery typically arises from which of the following?

A. RCA

B. Right marginal artery

C. Posterior interventricular artery

D. Anterior interventricular artery

E. Circumflex artery

16.2 In a balanced coronary artery pattern, the blood supply to the majority of the interventricular septum is derived from which of the following?

A. RCA

B. Internal mammary artery

C. Posterior interventricular artery

D. Anterior interventricular artery

E. Circumflex artery

16.3 As a cardiologist, you are concerned about blockage of the artery to the AV node in a patient. This artery typically arises from which of the following?

A. RCA

B. Right marginal artery

C. Posterior interventricular artery

D. Anterior interventricular artery

E. Circumflex artery

16.4 A 56-year-old man is complaining of chest pain that radiates to the jaw and left arm. A thallium stress test shows decreased perfusion to the heart that overlies the diaphragm. Which of the following coronary arteries is most likely to be blocked?

   A. Left anterior descending

   B. Right coronary

   C. Left main artery

   D. Left circumflex

ANSWERS

16.1 **A.** The SA node is typically supplied by the RCA.

16.2 **D.** Usually, the anterior two-thirds of the interventricular septum is supplied by the AV artery, and the right and left bundle branches of the conduction system are generally supplied by the anterior interventricular artery.

16.3 **A.** The AV node is also supplied by the RCA.

16.4 **B.** The inferior portion of the heart is supplied by the right coronary artery.

ANATOMY PEARLS

▶ In a balanced coronary circulation as described above, the conduction system nodes of the heart (SA and AV nodes) are typically supplied by the RCA.

▶ In a balanced coronary circulation, the anastomoses between branches of the RCA and LCA occur at the posterior coronary and posterior interventricular grooves.

▶ Most cardiac veins drain into the coronary sinus, which opens into the right atrium adjacent to the opening of the IVC.

REFERENCES

Gilroy AM, MacPherson BR, Ross LM. *Atlas of Anatomy*, 2nd ed. New York, NY: Thieme Medical Publishers; 2012:96–97.

Moore KL, Dalley AF, Agur AMR. *Clinically Oriented Anatomy*, 7th ed. Baltimore, MD: Lippincott Williams & Wilkins; 2014:144–148, 154–157.

Netter FH. *Atlas of Human Anatomy*, 6th ed. Philadelphia, PA: Saunders; 2014: plates 215–216.

A 31-year-old woman with one healthy child presents with a 2-year history of an inability to conceive. She states that her menstrual periods began at age 12 and occur at regular 28-day intervals. A biphasic basal body temperature chart is recorded. She denies having any sexually transmitted disease, and a hysterosalpingogram shows patent uterine tubes and a normal uterine cavity. Her husband is 34 years old, and his semen analysis is normal. In the presence of several normal test results for infertility, a laparoscopic examination of the pelvic cavity is scheduled. The physician performing the procedure carefully places the trocar lateral to the rectus abdominis muscle and its sheath to avoid injury to a major artery.

▶ What artery is being avoided?
▶ What is the anatomical location of this structure?

ANSWER TO CASE 17:

Inferior Epigastric Artery

Summary: An infertile couple is being evaluated, and several tests for infertility have shown normal results. A laparoscopic examination of the pelvic cavity is performed to rule out the presence of endometriosis. The trocar is specifically placed lateral to the rectus abdominis muscle and its sheath to avoid a major artery.

- **Artery that is avoided:** Inferior epigastric artery
- **Anatomical location of this artery:** Posterior to the rectus abdominis within the rectus sheath

CLINICAL CORRELATION

The presence of several normal test results indicates a need to rule out endometriosis in this infertile couple. **Endometriosis** is defined as ectopic endometrial tissue outside the uterus, typically adherent to the pelvic peritoneum. This tissue responds to a woman's hormonal cycles in the same way that the lining of the uterus responds. Although the mechanism is not fully understood, endometriosis may cause infertility by inhibiting ovulation, producing adhesions, or interfering with fertilization. Laparoscopic examination of the pelvic cavity is indicated and, if possible at this time, ablation of the endometrial tissue.

APPROACH TO:

Anterior Abdominal Wall

OBJECTIVES

1. Be able to describe the arterial blood supply of the anterior abdominal wall
2. Be able to describe the relation of these vessels to the musculature of the anterior abdominal wall, including the rectus sheath

DEFINITIONS

BIPHASIC BASAL BODY TEMPERATURE CHART: Elevation of the oral temperature during the second half of the menstrual cycle, indicating that the patient has ovulated

HYSTEROSALPINGOGRAM: Radiologic study in which radiopaque dye is injected into the uterine lumen through a transcervical catheter to evaluate the uterine cavity or the patency of the uterine tubes

LAPAROSCOPY: Surgical technique to visualize the peritoneal cavity through a rigid telescopic instrument called a **laparoscope**

ENDOMETRIOSIS: Condition in which the lining tissue of the uterus, the endometrium, is far outside the uterus, typically in the pelvic cavity or on the abdominal wall

DISCUSSION

The **anterior abdominal** wall is composed of three paired flat muscles that, in general, arise from bony structures posteriorly and whose fibrous aponeuroses form the rectus sheath and meet to form the **linea alba.** These muscles are, from superficial to deep: the **external abdominal oblique, internal abdominal oblique, and transversus abdominis** (Figure 17-1). These muscles are supplied by segmental branches of the thoracic and abdominal aorta: the 10th, 11th, and 12th (subcostal) intercostal arteries and the 1st or 2nd lumbar arteries. These **arteries,** their companion **veins,** and the **nerves** supplying the muscles are all found **in the interval between the internal abdominal oblique and the transversus abdominis muscles,** known as the **neurovascular plane. Superficial and deep circumflex iliac arteries** arise from the **femoral and external iliac arteries,** respectively; course parallel to the inguinal ligament; and supply the inferior abdominal wall in the inguinal region. The superficial epigastric arteries lie in the superficial fascia between the umbilicus and the pubic bone. The internal thoracic arteries divide into two terminal branches: the superior epigastric and musculophrenic arteries (Figure 17-2).

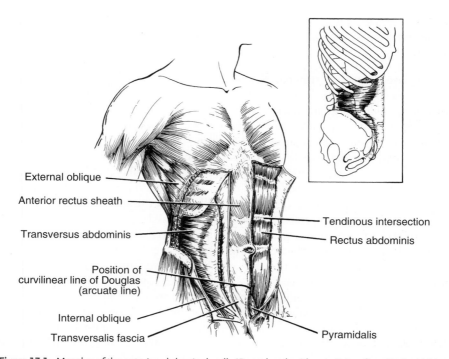

Figure 17-1. Muscles of the anterior abdominal wall. (*Reproduced, with permission, from Lindner HH. Clinical Anatomy. East Norwalk, CT: Appleton & Lange, 1989:291.*)

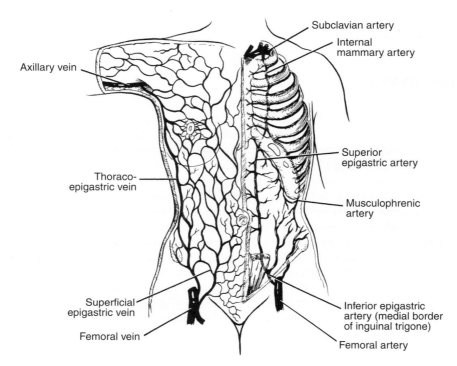

Figure 17-2. Arteries of the anterior abdominal wall. (*Reproduced, with permission, from Lindner HH. Clinical Anatomy. East Norwalk, CT: Appleton & Lange, 1989:299.*)

The anterior central region of the abdominal wall is formed by the paired **rectus abdominis** muscles, which attach to the pubic bone inferiorly and the rib cartilages superiorly, and lie just lateral to the linea alba. Each muscle is subdivided into short belly segments by typically three or more tendinous inscriptions. Each muscle is contained in a fibrous compartment, the **rectus sheath,** which is formed by the aponeuroses of the three flat abdominal muscles. Enclosing the superior three-fourths of each rectus muscle are anterior and posterior layers to the sheath. In this region, the aponeurosis of the internal abdominal muscle divides, and portions will pass anteriorly and posteriorly to the rectus muscle. It follows that the aponeuroses of the external oblique muscle and the transversus abdominis muscles **must pass anteriorly and posteriorly to the rectus, respectively.** At about the midpoint between the umbilicus and the pubic bone, the aponeuroses of all three flat abdominal muscles pass anterior to the rectus sheath; therefore, the inferior one-fourth of the rectus muscle has only an anterior rectus sheath. The inferior margin of the posterior rectus sheath can be identified in this region of transition as the **arcuate line.** The rectus sheath also contains the pyramidalis muscles, superior and inferior epigastric vessels, and terminations of the intercostal nerves that innervate the abdominal muscles. The superior one-fourth of the rectus muscle is supplied by the medial terminal branch of the **internal thoracic artery,** the **superior epigastric artery.** The **inferior epigastric artery** arises from the **external iliac artery** just before its exit from the abdomen to become the femoral artery. Each artery courses medially, external to the peritoneum, along the posterior surface of the rectus muscles. The superior and inferior epigastric arteries anastomose about halfway between the umbilicus and the xiphoid process.

COMPREHENSION QUESTIONS

17.1 A surgeon entering the abdominal cavity through the abdominal wall will take care to avoid injury to the vessels and nerves within the wall. The main portion of these vessels and nerves will be found immediately deep to which of the following?

A. Skin
B. Superficial fascia
C. External abdominal oblique muscle
D. Internal abdominal oblique muscle
E. Transversus abdominis muscle

17.2 As a surgeon performing an appendectomy, you encounter an artery and vein in the superficial fascia of the lower abdominal wall. These vessels are most likely which of the following?

A. Superficial epigastric artery and vein
B. Superficial circumflex iliac artery and vein
C. Intercostal artery and vein
D. Inferior epigastric artery and vein
E. Superior epigastric artery and vein

17.3 During surgery, you must incise the anterior rectus sheath between the xiphoid process and the umbilicus. In this region, the sheath is derived from the aponeurosis of which of the following?

A. External abdominal oblique only
B. Internal abdominal oblique only
C. External and internal abdominal oblique
D. Internal oblique and transversus abdominis
E. Transversus abdominis only

17.4 During a laparoscopic procedure, you observe the inferior epigastric vessels ascending on the posterior surface of the rectus abdominis muscle. They suddenly disappear from view by passing superior to which of the following?

A. Falx inguinalis
B. Linea semilunaris
C. Falciform ligament
D. Arcuate line
E. Transversalis fascia

ANSWERS

17.1 **D.** The main course of the intercostal vessels and nerves is deep to the internal abdominal oblique muscle in the neurovascular plane.

17.2 **A.** The superficial epigastric vessels lie within the superficial fascia.

17.3 **C.** The superior three-fourths of the anterior rectus sheath is derived from the aponeuroses of the external and internal abdominal oblique muscles.

17.4 **D.** As the inferior epigastric vessels ascend on the posterior surface of the rectus abdominis muscle, they will pass superiorly to the arcuate line, anteriorly to the posterior rectus sheath.

ANATOMY PEARLS

▶ The neurovascular plane of the anterolateral abdominal wall lies deep to the internal abdominal oblique muscle.

▶ Along the superior three-fourths of the rectus muscles, the internal oblique aponeurosis splits to contribute to the anterior and posterior rectus sheath layers.

▶ The inferior epigastric artery arises from the external iliac artery, lies on the posterior surface of the rectus muscle, and serves as its main blood supply.

REFERENCES

Gilroy AM, MacPherson BR, Ross LM. *Atlas of Anatomy,* 2nd ed. New York, NY: Thieme Medical Publishers; 2012:136−137, 146−147.

Moore KL, Dalley AF, Agur AMR. *Clinically Oriented Anatomy,* 7th ed. Baltimore, MD: Lippincott Williams & Wilkins; 2014:186−196, 198−199.

Netter FH. *Atlas of Human Anatomy,* 6th ed. Philadelphia, PA: Saunders; 2014: plates 247, 249, 251.

A 44-year-old man complains of discomfort in his right upper thigh for the past 6 months. He works in the garden department of a home improvement center. On examination, there is tenderness at the right inguinal area. When the patient performs a Valsalva maneuver (bearing down to increase intraabdominal pressure), a bulge appears superior to the inguinal crease near the pubic bone.

▶ What is the most likely diagnosis?
▶ What is the anatomical defect associated with this condition?

ANSWER TO CASE 18:
Inguinal Hernia

Summary: A 44-year-old man who works in the garden department of a home improvement center has a 6-month history of right groin pain. There is inguinal tenderness and a bulge following a Valsalva maneuver.

- **Most likely diagnosis:** Inguinal hernia
- **Associated anatomical defect:** Protrusion of an abdominal organ into the inguinal canal

CLINICAL CORRELATION

A **hernia** is defined as an abnormal protrusion of a structure through tissues that normally contain it. Inguinal hernias are the most common type of hernia, occurring in men and women, although they occur much more frequently in men. This patient's age and his occupation, which requires frequent lifting activity, suggest a direct or acquired inguinal hernia. Loss of tone in the musculature in the inguinal region predisposes to progressive stretching of the parietal peritoneum into the posterior inguinal canal with repeated increased intraabdominal pressure associated with the lifting activity. If the patient were a young man or child, an indirect or congenital inguinal hernia would be a more likely diagnosis. With an indirect hernia, the parietal peritoneum at the deep inguinal ring exists as a fingerlike protrusion into the inguinal canal. This is the result of faulty closure of the embryonic outpouching of peritoneum into the scrotum, called the **process vaginalis**. Indirect inguinal hernias enter the deep inguinal ring, stretch peritoneal tissue with repeated increases in intraabdominal pressure, traverse the length of the inguinal canal, and enter the scrotum. Surgical repair of the tissue defect is indicated to prevent incarceration, infarction, and necrosis of the herniated tissue, typically a loop of small intestine.

APPROACH TO:
The Inguinal Region

OBJECTIVES

1. Be able to describe the anatomy of the inguinal region
2. Be able to discern the anatomical basis for an indirect versus a direct inguinal hernial classification

DEFINITION

VALSALVA MANEUVER: Increase intraabdominal pressure by attempting to exhale with a closed glottis

DISCUSSION

The **inguinal region** is the junction between the **lower anterior abdominal** and the **upper anterior thigh.** It is the site at which several structures enter and exit the abdomen and therefore is an area of potential weakness in males and females. The **inguinal (Poupart) ligament** is an important anatomical structure and key landmark for this region. It is the thickened, rolled underedge of the inferior portion of the **external abdominal oblique aponeurosis.** It extends from the **anterior superior iliac spine** to the **pubic tubercle** and fuses inferiorly with the **fascia lata (deep fascia)** of the anterior thigh. At the pubic tubercle, the inguinal ligament continues postero-laterally on the **superior pubic ramus** (pectin of the pubic bone) as the **pectineal (Cooper) ligament.** At the point where these two ligaments are continuous and change directions, a ligamentous reflection fills the interval, forming the **lacunar (Gimbernat) ligament.**

The lacunar ligament forms a rigid medial margin for the femoral ring, leading to the femoral canal, the site for femoral hernias (Figure 18-1).

Although the external abdominal oblique muscle and aponeurosis constitute an essentially complete musculotendinous structure (except for the superficial inguinal ring), the **internal abdominal oblique** and **transversus abdominis muscles** are deficient because they originate from the iliopsoas fascia and arch medially to their **tendinous (falx inguinalis) insertions on the pubic tubercle** (Figure 18-2).

Structures enter and exit the abdomen superior to the inguinal ligament through an oblique passage known as the **inguinal canal.** The canal is frequently described as a tunnel, with openings, walls, floor, and so on. These boundary features are listed in Table 18-1.

Two points in Table 18-1 are of anatomical and clinical significance. First, as a result of the arching of the internal oblique and transversus abdominis muscles, the **posterior wall of the canal is deficient and weak,** as it is formed only by the **transversalis fascia** and parietal peritoneum. However, with increased intraabdominal pressure (as in lifting, a bowel movement, etc.), these muscles contract and descend

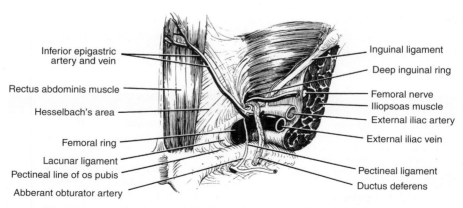

Inferior epigastric artery and vein

Rectus abdominis muscle

Hesselbach's area

Femoral ring

Lacunar ligament

Pectineal line of os pubis

Abberant obturator artery

Inguinal ligament

Deep inguinal ring

Femoral nerve

Iliopsoas muscle

External iliac artery

External iliac vein

Pectineal ligament

Ductus deferens

Figure 18-1. Inner surface of Hesselbach triangle. (*Reproduced, with permission, from Lindner HH. Clinical Anatomy. East Norwalk, CT: Appleton & Lange, 1989:288.*)

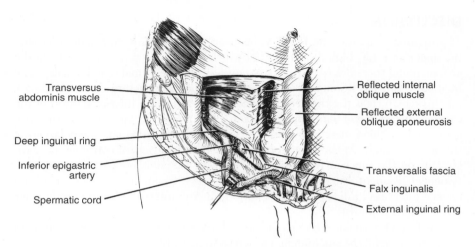

Figure 18-2. The ilioinguinal region. (*Reproduced, with permission, from Lindner HH. Clinical Anatomy. East Norwalk, CT: Appleton & Lange, 1989:291.*)

in a shutterlike fashion, thus reinforcing the posterior wall. Second, the outpouching of the transversalis fascia to form the **deep inguinal ring occurs immediately laterally to the inferior epigastric vessels** (see Case 17 for a discussion of their course). In addition, at the medial portion of the inguinal ligament on the interior of the abdominal wall, a clinically important **inguinal (Hesselbach) triangle** is formed by some of these structures. This triangle is formed by the **inguinal ligament, inferior epigastric vessels, and lateral margin of the rectus abdominis muscle** and corresponds to the area where the posterior wall of the canal is deficient because of the arching of the abdominal wall muscles described above.

In **females,** the **inguinal canal is traversed by the round ligament of the uterus;** in **males,** the **spermatic cord (ductus deferens and associated vessels and nerves)** passes through the canal. The **ilioinguinal nerve** is found in the canal in both sexes.

The inguinal region and canal serve as the site for inguinal hernias. Although hernias occur in both sexes, they are far more common in males. There are two types of inguinal hernias: indirect and direct. **Indirect or congenital inguinal hernias tend to occur in young males.** During embryonic descent of the testes, an outpouching

Table 18-1 • BOUNDARIES OF THE INGUINAL CANAL					
Anterior Wall	**Posterior Wall**	**Floor**	**Roof**	**External Opening**	**Internal Opening**
Aponeurosis of external abdominal oblique muscle	Transversalis fascia and parietal peritoneum	Inguinal ligament and lacunar ligament medially	Arching fibers internal oblique and transversus abdominis muscle	Superficial inguinal ring: triangular opening in external oblique aponeurosis	Site of outpouching of transversalis fascia: covered with parietal peritoneum

of **parietal peritoneum,** the **tunica vaginalis,** pushes through the lower abdominal wall, encountering first the transversalis fascia (thus forming the deep inguinal ring), slipping inferior to the transversus abdominis muscle, but catching the lower margin of the internal abdominal oblique muscle, and then pushing through the external abdominal oblique muscle **(forming the superficial inguinal ring).** The testes descend into the scrotum along the path created by the tunica vaginalis (and the gubernaculum). In normal development, this outpouching fuses and closes. If it does not fuse and close, a predisposing complete or partial path for the abnormal migration of an abdominal organ (usually small intestine) is established. The loop of small intestine would pass through the deep inguinal ring and the inguinal canal, and possibly through the superficial ring into the scrotum. By definition, **indirect inguinal hernias** leave the abdominal cavity **lateral** to the **inferior epigastric vessels** (through the deep inguinal ring).

Direct inguinal hernias are also called **acquired inguinal hernias** because they are seen in **older males** and are related to strenuous activity that increases intraabdominal pressure. It is believed that with aging there is loss of tone in the abdominal musculature, and the shutterlike actions described above for the internal abdominal oblique and transverses abdominis are diminished or lost. This predisposes abdominal organs to push directly anterior through the parietal peritoneum and transversalis fascia **in the inguinal triangle** area and into the posterior wall of the canal. Because of the larger herniation, these hernias tend not to enter the scrotum. **Direct inguinal hernias** by definition leave the abdomen **medial to the inferior epigastric vessels** because these vessels form the lateral boundary of the triangle.

COMPREHENSION QUESTIONS

18.1 As a physician examining the inguinal region of a patient, you note that the inguinal ligament will be a key landmark. This structure is a feature derived from which of the following?

    A. Superficial fascia

    B. Fascia lata of the thigh

    C. Aponeurosis of the external abdominal oblique

    D. Aponeurosis of the internal abdominal oblique

    E. Aponeurosis of the transversus abdominis

18.2 As you continue your examination to check for the presence of an inguinal hernia, you insert the tip of your finger into the superficial inguinal ring. This is an opening in which of the following?

    A. Superficial fascia

    B. Fascia lata of the thigh

    C. Aponeurosis of the external abdominal oblique

    D. Aponeurosis of the internal abdominal oblique

    E. Aponeurosis of the transversus abdominis

18.3 You are in the process of repairing a direct inguinal hernia. Which of the following anatomical relations will you find during surgery?

A. The hernia will enter the deep inguinal ring.

B. The hernia will enter the femoral ring.

C. The hernia will lie lateral to the inferior epigastric vessels.

D. The hernia will lie medial to the inferior epigastric vessels.

E. The hernia will lie inferior to the inguinal ligament.

18.4 A 58-year-old man who works in a warehouse lifting heavy boxes visits his physician complaining of pain in the groin area. On exam, he is found to have a large bulge superior to the right inguinal ligament. Imaging shows that the bulge arises medial to the inferior epigastric artery. This condition is most likely due to weakness of the

A. Inguinal ring

B. Femoral ring

C. Rectus abdominis muscle

D. Transversalis fascia

E. Cremasteric muscle

ANSWERS

18.1 **C.** The inguinal ligament is the inferior edge of the aponeurosis of the external abdominal oblique muscle.

18.2 **C.** The superficial inguinal ring is an opening in the aponeurosis of the external abdominal oblique muscle.

18.3 **D.** Direct inguinal hernias occur through the inguinal triangle, and the inferior epigastric vessels form the lateral boundary of this triangle. Hence, these vessels are lateral to the hernia.

18.4 **D.** Direct inguinal hernias usually present in older men, medial to the inferior epigastric vessels, and because of weakness of the transversalis fascia. Indirect hernias are present in male infants and are due to failure of the inguinal ring to close; the hernia is typically lateral to the inferior epigastric vessels.

ANATOMY PEARLS

▶ The external abdominal oblique aponeurosis forms the anterior wall and floor of the inguinal canal (inguinal ligament) and the superficial inguinal ring.

▶ The deep inguinal ring lies immediately lateral to the inferior epigastric vessels.

▶ Indirect inguinal hernias enter the deep inguinal ring (lateral to the epigastric vessels).

▶ Direct inguinal hernias enter the inguinal triangle (medial to the epigastric vessels).

REFERENCES

Gilroy AM, MacPherson BR, Ross LM. *Atlas of Anatomy*, 2nd ed. New York, NY: Thieme Medical Publishers; 2012:142–143, 147.

Moore KL, Dalley AF, Agur AMR. *Clinically Oriented Anatomy*, 7th ed. Baltimore, MD: Lippincott Williams & Wilkins; 2014:203–206, 212–214.

Netter FH. *Atlas of Human Anatomy*, 6th ed. Philadelphia, PA: Saunders; 2014: plates 245–247, 255–257.

A 42-year-old woman is seen by her primary care physician complaining of intermittent colicky pain. She describes the pain as being right upper quadrant (RUQ), starting shortly after eating a meal, and lasting about 30 min. During these episodes, she feels bloated and nauseated. The patient also states that over the past 2 days, her stools have become very light in color, like the color of sand, and her skin has become yellow.

▶ What is the most likely diagnosis?
▶ What is the anatomic basis for the clinical condition?

ANSWER TO CASE 19:

Gallstones

Summary: A 42-year-old woman presents with intermittent colicky RUQ abdominal pain shortly after eating, that lasts for about 30 min. It is associated with bloating, nausea, and a 2-day history of acholic stools and icterus.

- **Most likely diagnosis:** Gallstones
- **Anatomical basis for condition:** Bile duct obstruction, probably by gallstones

CLINICAL CORRELATION

This middle-aged woman has the typical symptoms of biliary colic, which is intermittent crampy abdominal pain in the epigastric region of the RUQ, sometimes radiating to the right shoulder. These symptoms typically appear after meals, particularly fatty meals. The more concerning signs are the light-colored stools (acholic) and jaundice (icterus). Gallstones (cholelithiasis) are precipitated bile salts in the gallbladder, which may produce inflammation of the gallbladder (cholecystitis). Stones can pass into the cystic duct and into the common bile duct. Since the common bile duct is formed by the union of the cystic and common hepatic ducts, obstruction of the common bile duct prevents bilirubin produced in the liver from reaching the small intestines. The stools thus lack this pigment. As a secondary result of the obstruction, serum bilirubin is elevated, and precipitates in the skin, resulting in the yellow tint. Ultrasound can often make the initial diagnosis. Removal of a common bile duct stone can be performed by upper GI endoscopy through the ampulla of Vater or surgically.

APPROACH TO:

The Gallbladder

OBJECTIVES

1. Be able to describe the anatomy of the gallbladder and hepatobiliary duct system
2. Be able to describe the clinically important anatomical relationships of the cystic and common bile ducts

DEFINITION

CHOLECYSTITIS: Inflammation of the gallbladder often associated with gallstones

DISCUSSION

The **gallbladder** is an inverted, pear-shaped, fibromuscular sac that provides temporary storage and intermittent release of bile, which is produced in the liver. Its surface position can be approximated at the intersection of the right margin of the

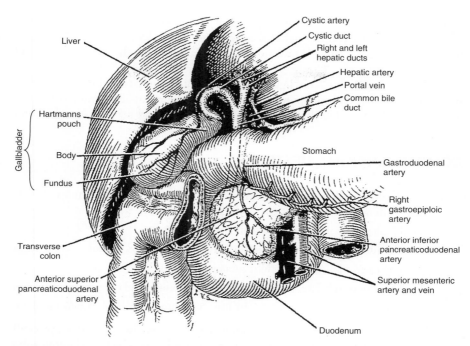

Figure 19-1. Relationships of the gallbladder. (*Reproduced, with permission, from Lindner HH. Clinical Anatomy. East Norwalk, CT: Appleton & Lange, 1989.*)

rectus sheath (linea semilunaris) and the right costal margin. Its anterior surface is fused to the liver between the right and quadrate lobes, and its fundus, lateral, and posterior surfaces are covered with visceral peritoneum. It is anatomically divided into fundus, body, and neck, which is continuous with the **cystic duct.** The mucosa of the neck and cystic duct are spiral fold, which acts as a valve to keep the lumen of the duct and neck open to receive bile. The gallbladder and cystic duct are supplied by the cystic artery, typically a branch of the right hepatic artery (see Figure 19-1).

The biliary duct system begins as **bile canaliculi** between hepatocytes within the liver. The canaliculi empty into microscopic interlobular bile ducts, which unite to form increasingly large ducts, eventually forming segmental and lobar ducts draining anatomical subdivisions of the liver of the same name. Ultimately, right and left hepatic ducts emerge from the liver's porta hepatis and unite to form the **common hepatic duct** within the hepatoduodenal ligament (a portion of the lesser omentum). The cystic duct joins the common hepatic duct from the right to form the **common bile duct**, which passes inferior within the hepatoduodenal ligament and then passes posterior to the first part of the duodenum. It turns slightly to the right on or within the posterior surface of the pancreas. As it approaches the posteromedial wall of the duodenum, it is typically joined by the **main pancreatic duct** to form the hepatopancreatic ampulla, which opens on the major duodenal papilla.

At **the porta hepatis**, the right and left hepatic ducts are the most anterior structures. The hepatic arteries (right and left) lie posterior to the hepatic ducts, and the branches of the portal vein lie most posterior. The common hepatic duct (on the left), the cystic duct (on the right), and the inferior border of the liver (superior) form the cystohepatic **triangle of Calot**, which contains the right hepatic artery and its cystic artery branch.

Within the **hepatoduodenal ligament**, the anterior boundary of the epiploic foramen (of Winslow), the common bile duct lies to the right, the common hepatic artery lies to the left, and the portal vein lies posterior and between the duct and the artery.

COMPREHENSION QUESTIONS

19.1 Which of the following is the correct landmark for locating the normal position of the gallbladder during a physical examination?

    A. The lowest point of the left subcostal margin

    B. The junction between the left linea semilunaris and the subcostal margin

    C. The lowest point of the right subcostal margin

    D. The junction between the right linea semilunaris and the subcostal margin

    E. The junction between the right linea semilunaris and the subcostal plane

19.2 During a surgical procedure in which you will remove the gallbladder, you will expect its blood supply, the cystic artery, to arise from which of the following arteries?

    A. Right hepatic artery

    B. Left hepatic artery

    C. Proper hepatic artery

    D. Common hepatic artery

    E. Right gastric artery

19.3 During the surgical procedure described in question 19.2, your index finger is placed into the epiploic foramen. Which of the following structures would be inferior to your finger?

    A. Caudate lobe of the liver

    B. First part of the duodenum

    C. Inferior vena cava

    D. Portal vein

    E. Hepatic artery

19.4 A 45-year-old woman with a history of gallstones visits the emergency department complaining of severe abdominal pain and vomiting for 1 day. The exam shows a distended abdomen and high-pitched bowel sounds. Radiographs of the abdomen show air in the biliary tree and the gallbladder. Which of the following is the most likely location for the gallstone to be found?

A. Common bile duct

B. Duodenum

C. Sphincter of Oddi

D. Jejunum

E. Ileum

F. Ascending colon

ANSWERS

19.1 **D.** The gallbladder is normally located at the junction between the right semilunar line and the right subcostal margin.

19.2 **A.** The cystic artery typically is a branch of the right hepatic artery.

19.3 **B.** The first part of the duodenum will lie inferior to a finger within the epiploic foramen.

19.4 **E.** The diagnosis is likely gallstone ileus, in which a large gallstone is impacted in the ileocecal valve. Air in the biliary tree is caused by a fistula between the bowel and the biliary tree, allowing air from the bowel to enter the biliary system. The gallstone causes bowel obstruction. This is a surgical emergency.

ANATOMY PEARLS

► The gallbladder fossa lies between the right and quadrate lobes of the liver.

► The cystic artery is usually a branch of the right hepatic artery.

► The hepatic ducts are the most anterior structures at the porta hepatis.

► The bile duct lie to the right within the hepatoduodenal ligament.

REFERENCES

Gilroy AM, MacPherson BR, Ross LM. *Atlas of Anatomy*, 2nd ed. New York, NY: Thieme Medical Publishers; 2012:168–169.

Moore KL, Dalley AF, Agur AMR. *Clinically Oriented Anatomy*, 7th ed. Baltimore, MD: Lippincott Williams & Wilkins; 2014:277–280, 286–288.

Netter FH. *Atlas of Human Anatomy*, 6th ed. Philadelphia, PA: Saunders; 2014: plates 280–281, 283.

A 62-year-old woman complains of the sudden onset of severe midabdominal pain that has been increasing over the past 3 h. She has a history of myocardial ischemia and peripheral vascular disease. The patient states that she has had nausea and vomiting. On examination, she is writhing in pain. Her abdomen has normal bowel sounds and minimal tenderness. A small amount of blood is present in the stool specimen. The electrolytes show a low bicarbonate level at 15 mEq/L, and the serum lactate level is high, which are indicative of tissue receiving insufficient oxygenation leading to tissue injury. A surgeon who is concerned about intestinal ischemia has been called to evaluate the patient.

▶ What is the most likely diagnosis?
▶ What anatomical structure is likely involved?

ANSWER TO CASE 20:

Small Bowel Mesenteric Angina

Summary: A 62-year-old woman with widespread atherosclerotic vascular disease complains of a 3-h history of severe midabdominal pain accompanied by nausea and vomiting. Although she is writhing in pain, her bowel sounds are normal, and there is minimal tenderness. Blood is present in the stool, and electrolytes show low levels of bicarbonate at 15 mEq/L and high levels of lactate; these findings are attributed to a lack of oxygen to intestinal tissue, leading to anaerobic metabolism. The surgeon is concerned about ischemia.

- **Most likely diagnosis:** Mesenteric ischemia

- **Anatomical structures likely involved:** Arteries that supply the small bowel, probably branches of the superior mesenteric artery (SMA)

CLINICAL CORRELATION

This elderly woman complains of sudden-onset severe midabdominal pain that is inconsistent with the physical findings. She has a history of widespread atherosclerotic vascular disease affecting the coronary arteries and peripheral vasculature. The presence of blood in the stool suggests bowel injury, and the low level of serum bicarbonate is consistent with a metabolic acidemia. Bowel ischemia or necrosis is causative. Arterial occlusion may occur from rupture of the atherosclerotic plaque or embolization from another clot. This patient's midabdominal symptoms suggest arteriography of the SMA, and the celiac artery might be diagnostic. On confirmation, surgical embolectomy is usually helpful. The mortality rate is high in such patients.

Although the first part of the duodenum is supplied by the superior pancreatico-duodenal artery, which receives its blood from the celiac artery, the remainder of the small intestines is supplied by branches of the SMA.

APPROACH TO:

Vascular Supply to the Bowel

OBJECTIVES

1. Be able to describe the general plan for the arterial blood supply to the abdominal viscera

2. Be able to describe the anatomy and distribution of the SMA

DEFINITIONS

ATHEROSCLEROTIC VASCULAR DISEASE: Disease in which deposits of plaques of cholesterol and lipid form within the intima of small and medium arteries

ANGINA: Pain, often severe, due to decreased blood flow to an organ such as the heart or intestines

SUPERIOR MESENTERIC ARTERY: Unpaired arterial branch of the abdominal aorta that supplies portions of the duodenum, jejunum, ileum, cecum, appendix, ascending colon, and most of the transverse colon

DISCUSSION

The **abdominal gastrointestinal viscera** are supplied by the **three major unpaired branches of the abdominal aorta: celiac artery (trunk), SMA, and inferior mesenteric artery (IMA).** These three arteries supply organs embryologically derived from the **foregut, midgut, and hindgut,** respectively.

The **duodenum** proximal to the entrance of the common bile duct receives its blood supply from the **superior pancreaticoduodenal artery,** a branch of the gastroduodenal artery from the celiac artery. The remainder of the small intestines is supplied by the **SMA** (Figure 20-1). The SMA arises from the abdominal aorta at the level of the lower border of L1, posterior to the neck of the pancreas. As it emerges from behind the pancreas, it passes anterior to the uncinate process of the pancreas and the third part of the duodenum and enters the **root of the mesentery.** As it enters the mesenteric root, it gives off its **inferior pancreaticoduodenal** and **middle colic arteries,** the latter to the transverse colon within its mesentery, the **transverse mesocolon.**

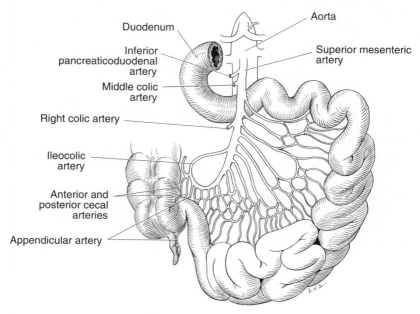

Figure 20-1. Superior mesenteric arterial supply to the small bowel. (*Reproduced, with permission, from Lindner HH. Clinical Anatomy. East Norwalk, CT: Appleton & Lange, 1989:353.*)

As the SMA descends toward the ileocolic junction, **15 to 18 intestinal branches** arise, which pass between the layers of the mesentery, and are united by increasingly complex anatomical **arcades.** The arcades closest to the mesenteric attachment to the jejunum and ileum give off increasingly shorter straight arteries (vasa recta) that enter the small intestines. Other branches of the SMA include the **right colic** to the ascending colon and the **ileocolic** to the cecum, appendix, and ascending colon.

COMPREHENSION QUESTIONS

20.1 During a surgical procedure, you have elevated the transverse colon and note an artery in the transverse mesocolon. What is this vessel?
   A. Right gastroomental (*gastroepiploic*) artery
   B. Middle colic artery
   C. Inferior pancreaticoduodenal artery
   D. Right colic artery
   E. Left colic artery

20.2 During surgery you note a retroperitoneal artery crossing the right side of the posterior abdominal wall and supplying the ascending colon. Which vessel is this?
   A. Middle colic artery
   B. Left colic artery
   C. Ileocolic artery
   D. Right colic artery
   E. Sigmoidal artery

20.3 A 44-year-old accountant develops a bleeding ulcer around tax time. The gastroenterologist visualizes the ulcer in the proximal duodenum. A radiologist has been called to cannulate and embolize the artery supplying the ulcer. Which of the following arteries does the radiologist need to cannulate?
   A. Celiac artery
   B. SMA
   C. IMA
   D. Superior epigastric artery

ANSWERS

20.1 **B.** The middle colic artery courses through the transverse mesocolon to supply the transverse colon.

20.2 **D.** The right colic artery supplies the ascending colon and is retroperitoneal.

20.3 **A.** The superior pancreaticoduodenal artery is a terminal branch that arises from the celiac artery.

ANATOMY PEARLS

► The SMA arises from the aorta opposite L1 posteriorly to the neck of the pancreas but crosses anterior to the third part of the duodenum.

► The celiac artery and SMA anastomose with each other through the pancreaticoduodenal arteries.

► The SMA intestinal arcades increase in complexity, but the vasa recta decrease in length from proximal to distal.

REFERENCES

Gilroy AM, MacPherson BR, Ross LM. *Atlas of Anatomy*, 2nd ed. New York, NY: Thieme Medical Publishers; 2012:190−191.

Moore KL, Dalley AF, Agur AMR. *Clinically Oriented Anatomy*, 7th ed. Baltimore, MD: Lippincott Williams & Wilkins; 2014:226−228, 243−246.

Netter FH. *Atlas of Human Anatomy*, 6th ed. Philadelphia, PA: Saunders; 2014: plates 287−288.

An 18-year-old male college student complains of 12-h abdominal pain that began around his umbilicus but then shifted to the right lower quadrant (RLQ) and right side. He indicates that he has been nauseous over the past several hours. His temperature is 99.4°F. On physical examination, there is mild abdominal tenderness, particularly in the RLQ, but also on the right side. The laboratory analysis of the urine is normal.

▶ What is the most likely diagnosis?
▶ What accounts for the shift in location of the pain?

ANSWER TO CASE 21:

Acute Appendicitis

Summary: An 18-year-old man complains of 12-h abdominal pain that is initially periumbilical and then migrates to the RLQ. He has some nausea and a low-grade fever. The abdomen is tender in the RLQ and right lateral region. The urinalysis is normal.

- **Most likely diagnosis:** Appendicitis, possibly retrocecal.

- **Cause of shift in location of pain:** Pain initially irritates the visceral peritoneum, is referred to the periumbilical area, and then localizes to RLQ as the appendicitis worsens and inflames the parietal peritoneum.

CLINICAL CORRELATION

This college student's complaints are suspicious for appendicitis. The appendix is a small diverticulum that arises from the cecum and is typically free in the peritoneal cavity. Not infrequently, however, it is retrocecal in location and causes right-side or flank tenderness and very few peritoneal signs. Initially, the abdominal pain is vaguely and generally located to the periumbilical region, but with time, it becomes sharper and precisely located to the RLQ. Nausea is common but presents after the onset of pain. Men and women are equally affected by appendicitis, but the diagnosis is usually more straightforward in men. A serum leukocyte count may be helpful. Ultimately, the suspicion is a clinical one, and diagnostic laparoscopy is undertaken to visualize the appendix. If appendicitis is confirmed, surgery is indicated.

APPROACH TO:

The Large Bowel

OBJECTIVES

1. Be able to describe the anatomy of the appendix and large intestine

2. Be able to describe the mechanism for referred pain

3. Be able to describe the general anatomic pattern for abdominal pain

DEFINITIONS

APPENDICITIS: Inflammation of the appendix that is often associated with a fecalith, a small piece of stool that occludes the proximal appendix

REFERRED PAIN: Pain that originates from a deep structure that is perceived at the surface of the body, often at a different location

DISCUSSION

The typical position of the appendix can be approximated at a **point (McBurney)** one-third of the way along a line drawn from the right anterior superior iliac spine to the umbilicus. The **appendix is an elongated diverticulum** that arises from the **cecum** inferior to the ileocecal junction (Figure 21-1). The three longitudinal smooth muscle bands characteristic of the cecum and colon, the **teniae coli,** can be traced inferiorly to the posteromedial origin of the appendix from the cecum. The appendix lies in the margin of a small triangular mesentery, the mesoappendix, within which the **appendicular artery** (a branch of the ileocolic artery) is also found. The posterior surface of the cecum is often covered with visceral peritoneum, creating a **retrocecal recess.** In close to 66 percent of individuals, the appendix is retrocecal in position and is found in this recess. In almost 33 percent of individuals, the appendix is free and extends inferiorly toward or over the pelvic brim. The cecum and the appendix can lie at higher or lower positions relative to the McBurney point as a result of faulty embryonic gut rotation.

The **large intestines** are characterized by the presence of **teniae coli, haustra, omental appendices, and their large diameter.** The **cecum** is the pouchlike first part of the large intestines into which the ileum opens and the appendix arises. It is continuous superiorly with the **ascending colon,** which is the shortest segment of colon, is retroperitoneal (lacks a mesentery), is continuous with the transverse colon at the right colic (hepatic) flexure, and is supplied by the ileocolic and right colic branches of the superior mesenteric artery (SMA). The **transverse colon** is the longest segment of colon, begins at the right colic flexure, and is continuous with the descending colon at the more superiorly positioned left colic (splenic) flexure.

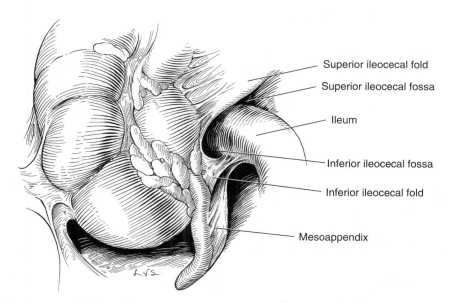

Superior ileocecal fold

Superior ileocecal fossa

Ileum

Inferior ileocecal fossa

Inferior ileocecal fold

Mesoappendix

Figure 21-1. Cecal folds and fossae. (*Reproduced, with permission, from Lindner HH. Clinical Anatomy. East Norwalk, CT: Appleton & Lange, 1989:361.*)

It is intraperitoneal, as it is suspended by its mesentery, the transverse mesocolon. The **middle colic artery branch of the SMA** lies within the mesentery. The **descending colon** is retroperitoneal, continuous with the sigmoid colon near the left iliac crest, and is supplied by the **left colic artery,** a branch of the **inferior mesenteric artery (IMA).** The **sigmoid colon** is suspended by its mesentery, the sigmoid mesocolon, in which its blood supply, the several sigmoidal arteries, are found. The sigmoid colon ends at the rectosigmoid junction, which lies at the S3 vertebral level. The arteries that supply the colon are connected by continuous arterial anastomoses called the **marginal arteries.**

The **initial vague, poorly localized pain** of appendicitis results from stretching of the **visceral peritoneum** secondary to inflammation of the organ. The cell bodies of the visceral afferent nerve fibers from the appendix lie in the dorsal root ganglia and enter the spinal cord at levels T8 through T10. Sensory fibers from the umbilicus enter the spinal cord at T10. The brain misinterprets (refers) the pain from the appendix as arising from the umbilical and nearby abdominal wall. This is called **referred pain.** As the inflammatory process progresses, adjacent **parietal peritoneum** is typically irritated, and the pain shifts to the actual location of the appendix in the **RLQ.** The parietal peritoneum is innervated by somatic sensory nerve fibers and, when irritated, produces sharp, well-localized pain sensation. If the appendix is **retrocecal,** the parietal peritoneum of the posterior abdominal wall is irritated, resulting in **side or flank tenderness.**

Pain originating from **foregut**-derived organs and supplied by the **celiac artery** is generally perceived in the **epigastric region.** Pain from **midgut**-derived organs supplied by the **SMA** is perceived in the **periumbilical region,** and pain in the **infraumbilical** region arises from **hindgut organs (IMA).**

COMPREHENSION QUESTIONS

21.1 You are at surgery for the removal of a suspected appendicitis, but the appendix is not visible. The appendix is likely to be which of the following?

A. Anticecal

B. Paracecal

C. Paracolic

D. Retrocecal

E. Retrocolic

21.2 Which of the following techniques could you use to precisely locate the appendix?

A. Locate a region devoid of haustra

B. Trace the right collect artery

C. Trace the ileocolic artery

D. Trace the teniae (*taeniae*) coli on the cecum

E. Examine the pelvic cavity

21.3 A patient with infraumbilical pain is likely to have a disorder of which organ?

  A. Appendix

  B. Ascending colon

  C. Ileum

  D. Stomach

  E. Sigmoid colon

ANSWERS

21.1 **D.** The appendix is retrocecal in position in almost 66 percent of the population.

21.2 **D.** The three teniae coli converge at the base of the appendix on the cecum.

21.3 **E.** Infraumbilical pain typically arises from hindgut-derived structures such as the sigmoid colon.

ANATOMY PEARLS

▶ The appendix typically lies at the McBurney point and is retrocecal in about 66 percent of the population.

▶ The SMA and IMA anastomose with each other through the marginal artery.

▶ The initial referred pain of appendicitis is to the periumbilical region.

REFERENCES

Gilroy AM, MacPherson BR, Ross LM. *Atlas of Anatomy,* 2nd ed. New York, NY: Thieme Medical Publishers; 2012:162–163.

Moore KL, Dalley AF, Agur AMR. *Clinically Oriented Anatomy,* 7th ed. Baltimore, MD: Lippincott Williams & Wilkins; 2014:247–249, 259–260.

Netter FH. *Atlas of Human Anatomy,* 6th ed. Philadelphia, PA: Saunders; 2014: plates 273–276, 303.

A 30-year-old man is admitted to the hospital for severe constant abdominal pain with nausea and vomiting since the previous day. He states that the pain radiates straight to his back and feels "like it's boring a hole right through me from front to back." He reports no other medical problems, but drinks one or two 6-packs of beer each weekend. He denies having diarrhea or fever. The serum amylase and lipase levels are markedly elevated.

► What is the most likely diagnosis?
► What is the anatomical location of the structure involved?

ANSWER TO CASE 22:

Pancreatitis

Summary: A 30-year-old man who drinks alcohol is admitted to the hospital for severe abdominal pain with nausea and vomiting for 24-h duration. He states that the pain radiates straight to his back. The serum amylase and lipase levels are markedly elevated.

- **Most likely diagnosis:** Acute pancreatitis

- **Anatomical location of the structure affected:** Retroperitoneal, posterior to the stomach and the lesser peritoneal sac (omental bursa)

CLINICAL CORRELATION

The pancreas is a retroperitoneal organ, posterior to the stomach and lesser sac, partly surrounded by the duodenum. It is an exocrine gland that secretes digestive enzymes and an endocrine gland that produces insulin and glucagon to regulate blood glucose levels. Noninfectious inflammation of the pancreas is most commonly caused by alcohol abuse or gallstones. The inflammation is secondary to autodigestion of the pancreatic tissue by the exocrine secretions. Marked vomiting is typical, and serum amylase or lipase levels are elevated. Immediate management includes restricting oral intake, monitoring fluid and electrolyte balance, and pain control. The pancreatitis sometimes may be so severe as to produce hemorrhage into the pancreas or pulmonary injury. These complications are associated with higher mortality rates.

APPROACH TO:

The Pancreas

OBJECTIVES

1. Be able to describe the anatomy of the pancreas and its relations to the duodenum and spleen

2. Be able to describe the retroperitoneal relations of the pancreas

DEFINITIONS

PANCREATITIS: Inflammation of the pancreas

RETROPERITONEAL: Posterior or external to the peritoneal cavity

OMENTAL BURSA: Subdivision of the peritoneal cavity posterior to the stomach and lesser omentum

DISCUSSION

The **pancreas** is a **retroperitoneal gland** that is **exocrine** (secretes digestive enzymes released into the duodenum) and **endocrine** (source of insulin and glucagon released into the bloodstream). It lies posterior to the omental bursa (lesser sac). The gland is anatomically divided into **head, neck, body,** and **tail** regions and is diagonally placed across the posterior abdominal wall (Figure 22-1). The head of the pancreas lies within the curve of the second and third parts of the duodenum, and its inferior portion forms a hooklike uncinate process that lies posterior to the superior mesenteric vessels. The neck lies at the L1 vertebral level, with the pylorus of the stomach immediately superior. The portal vein is formed posteriorly by the union of the splenic vein and superior mesenteric vein (SMV). The body of the gland passes superiorly to the left, with the tortuous splenic artery along its superior border. The short tail of the pancreas lies within the splenorenal ligament and may contact the hilum of the spleen (Table 22-1).

The exocrine pancreas is drained by a **main pancreatic duct,** which begins in the tail and passes to the right through the body, neck, and inferior portion of the head. The **duct pierces the wall of the second part of the duodenum** in close association with the **common bile duct,** with which it typically unites to form the **hepatopancreatic ampulla,** which, in turn, opens through the major **duodenal papilla.** Several smooth muscle sphincters surround these ducts, which may enter the duodenum separately at the papilla. The superior portion of the head is drained by an accessory pancreatic duct that usually joins the main duct but may drain separately into the duodenum at the minor duodenal papilla. The head of the pancreas receives

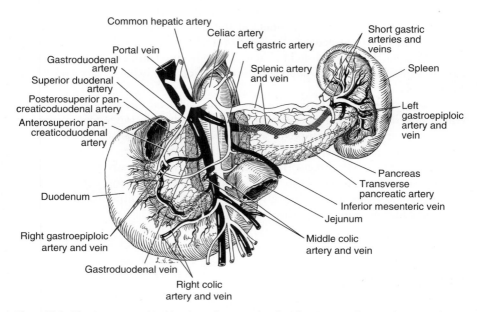

Figure 22-1. The pancreas and its blood supply. (*Reproduced, with permission, from Lindner HH. Clinical Anatomy. East Norwalk, CT: Appleton & Lange, 1989:346.*)

Table 22-1 • STRUCTURES POSTERIOR TO THE PANCREAS			
Head	Neck	Body	Tail
IVC	SMA and SMV	Aorta	Splenic vein
Right renal vessels	Splenic vein	SMA	
Left renal vein	Portal vein (formed)	Splenic vein	
		Left kidney and suprarenal gland	
		Left renal vessels	

its arterial blood supply primarily from superior and inferior **pancreaticoduodenal arteries** from the celiac and superior mesenteric artery (SMA), respectively, whereas the neck, body, and tail receive branches from the splenic artery.

The **duodenum** is the first, shortest, widest, and least mobile portion of the small intestine. It is anatomically subdivided into four parts, and its C-shaped configuration is intimately related to the pancreas. The superior or first part is the posteriorly directed continuation of the pylorus of the stomach, and it lies at the L1 vertebral level. Its first portion or ampulla (clinically, the duodenal cap) is intraperitoneal, within the hepatoduodenal ligament. The remainder is retroperitoneal. The descending or second part is retroperitoneal, lies opposite L1 through L3, and receives the pancreatic and bile ducts (hepatopancreatic ampulla) at the major duodenal papilla on its posteromedial wall. The horizontal or third part is also retroperitoneal, passes to the left, and crosses L3. The SMA and SMV cross this part of the duodenum anteriorly. The ascending or fourth part lies on the left side of the L3 and L2 vertebrae and is retroperitoneal, except perhaps for the last few millimeters as it becomes continuous with the jejunum at the duodenojejunal junction, indicated anatomically by the **suspensory ligament of Treitz.** The clinically important relations of the duodenum are listed in Table 22-2. The **duodenum** is supplied by **superior and inferior pancreaticoduodenal arteries** from the **celiac artery and SMA, respectively.**

The **spleen** is the largest lymph organ of the body and functions as if it were a lymph node for the circulatory system. It is intraperitoneal, suspended in the left upper quadrant by the **gastrosplenic and splenorenal ligaments** (subdivisions of the greater omentum). It lies parallel to the 10th rib and overlaps the 9th and 11th ribs. It has a convex diaphragmatic surface and concave hilum, where the ligaments attach. The **splenic artery** (a major branch of the celiac artery) enters, and the splenic vein exits the spleen through the hilum and is within the splenorenal ligament, in addition to the tail of the pancreas.

Table 22-2 • ANATOMICAL RELATIONS OF THE DUODENUM				
	Anterior	**Posterior**	**Medial**	**Superior**
Superior or first part	Gallbladder Quadrate lobe of liver	Bile duct Gastroduodenal artery Portal vein IVC		Epiploic foramen
Descending or second part	Transverse mesocolon Transverse colon Small intestines	Hilum right kidney Renal vessels and pelvis Right ureter Right psoas muscle	Head of pancreas Bile and pancreatic ducts	
Horizontal or third part	SMA and SMV Small intestines	IVC and aorta Right ureter Right psoas muscle		Head and uncinate process of pancreas SMA and SMV
Ascending or fourth part	Root of mesentery	Aorta, left side Left psoas muscle	Head of pancreas	Body of pancreas

COMPREHENSION QUESTIONS

22.1 You are at surgery and are about to mobilize the second portion of the duodenum and the head of the pancreas. You note an artery and vein passing anteriorly to the uncinate process of the pancreas and the third portion of the duodenum. Which vessels are these?

A. SMA and SMV

B. Inferior mesenteric artery and vein

C. Gastroduodenal artery and vein

D. Superior pancreaticoduodenal artery and vein

E. Middle colic artery and vein

22.2 As you proceed to elevate the duodenum and pancreas, you note two veins posterior to the neck of the pancreas uniting to form a large vein that passes superiorly. Which large vein has been formed?

A. Splenic vein

B. IVC

C. Portal vein

D. Right gastric vein

E. Middle colic vein

22.3 As you continue, you also note a large, tortuous artery passing to the left along the superior border of the pancreas. This is likely to be which of the following?

A. Left renal artery

B. IMA

C. Splenic artery

D. Left gastroomental (*gastroepiploic*) artery

E. Left colic artery

22.4 A 34-year-old man is involved in a motor vehicle accident. He is brought into the emergency room and is noted to have a hematoma involving the pancreas. What is the most likely location of this hematoma?

A. Intraperitoneal midline

B. Intraperitonal right side

C. Intraperitoneal left side

D. Retroperitoneal midline

E. Retroperitoneal behind spleen

ANSWERS

22.1 **A.** The SMA and SMV emerge from between the head and uncinate process of the pancreas to cross the uncinate process and the third portion of the duodenum.

22.2 **C.** The portal vein is formed by the union of the superior mesenteric and splenic veins posterior to the neck of the pancreas.

22.3 **C.** The splenic artery, the most tortuous artery of the body, is located along the superior border of the pancreas as it passes to the left toward the spleen.

22.4 **D.** Abdominal and pelvic blunt-force trauma such as a motor vehicle accident is commonly associated with retroperitoneal hematoma, such as involving the pancreas. The pancreas is located in the retroperitoneal space, and hematomas are typically in the midline. CT imaging can be used to identify these injuries.

ANATOMY PEARLS

▶ The pancreas is retroperitoneal, posterior to the omental bursa.

▶ The splenic artery passes along the superior border of the pancreas, whereas the splenic vein lies posterior.

▶ The portal vein is formed posterior to the neck of the pancreas.

▶ The hepatoduodenal papilla (bile and pancreatic ducts) opens onto the major duodenal papilla on the posteromedial wall of the second part of the duodenum.

▶ The second part of the duodenum is related to the right kidney hilum, pelvis and ureter, and renal vessels posteriorly.

▶ The third part of the duodenum is crossed anteriorly by the SMA and SMV.

REFERENCES

Gilroy AM, MacPherson BR, Ross LM. *Atlas of Anatomy*, 2nd ed. New York, NY: Thieme Medical Publishers; 2012:170−177.

Moore KL, Dalley AF, Agur AMR. *Clinically Oriented Anatomy*, 7th ed. Baltimore, MD: Lippincott Williams & Wilkins; 2014:265−268, 282−283.

Netter FH. *Atlas of Human Anatomy*, 6th ed. Philadelphia, PA: Saunders; 2014: plates 281, 284, 286−287, 289, 294.

A 38-year-old man comes into the emergency department presenting with fatigue and abdominal swelling. For several months, he has noticed that his abdomen has been growing larger and that his skin has turned yellow. He denies any medical problems but admits to drinking alcohol almost every day. On examination, his skin clearly has a yellow hue indicative of icterus. His palms have some redness. His abdomen is markedly distended and tense, and a fluid wave is present. Prominent vascular markings appear on the surface of the abdomen.

▶ What is the most likely diagnosis?
▶ What organs are likely to be affected?

ANSWER TO CASE 23:

Cirrhosis

Summary: A 38-year-old icteric man who abuses alcohol enters the emergency department for fatigue and abdominal "swelling." He has palmar erythema and abdominal distention with a positive fluid wave and prominent vascular markings.

- **Most likely diagnosis:** Alcoholic cirrhosis with portal hypertension
- **Organs likely affected:** Liver and those drained by the portal venous system

CLINICAL CORRELATION

This patient abuses alcohol and has manifestations of end-stage liver disease (cirrhosis). Cirrhosis results in severe fibrotic scarring of the liver, which decreases blood flow through the organ. Hypertension in the portal venous system is the result, with collateral venous flow, especially in organs having venous drainage by the portal and vena caval systems, such as the abdominal surface, and the esophagus. The spleen is frequently enlarged, and the ascites, fluid within the peritoneal cavity, is due to liver insufficiency. Death may ensue as a result of bleeding from esophageal varices or bacterial peritonitis of the ascitic fluid. Marked hepatic insufficiency is another complication.

APPROACH TO:

The Liver

OBJECTIVES

1. Be able to describe the anatomy of the liver and its unique blood supply
2. Be able to sketch the anatomy of the portal venous system and the clinically important sites of anastomosis with the vena caval system

DEFINITIONS

CIRRHOSIS: Disease of progressive degeneration of the liver in which damage to the liver cells results in nodular regeneration, fibrosis, and impedance

PORTACAVAL ANASTOMOSIS: Communication between tributaries of the portal venous system and the systemic venous system

PORTAL HYPERTENSION: Increased pressure in the portal venous system with resultant reverse flow, typically due to obstructed venous flow through the liver, as in cirrhosis

FLUID WAVE: A maneuver during physical examination in which tapping on one side of the abdomen leads to the sensation of a force traveling to the other side of the abdomen, suggesting the presence of intraabdominal fluid

DISCUSSION

The **liver,** the **largest internal organ,** has a **convex diaphragmatic surface** that conforms to the curvature of the diaphragm and an **irregular concave visceral surface.** The liver is covered with visceral peritoneum over most of its surface and is suspended by several mesenteric structures called **ligaments.** The **falciform ligament** (with the **round ligament of the liver,** the adult remnant of the umbilical vein, in its free margin) is reflected onto the anterior abdominal wall and divides the liver into apparent right and left anatomical lobes. As the falciform ligament passes onto the superior surface of the liver, the two layers of peritoneum diverge to the right and to the left, creating the anterior layers of the **coronary ligaments.** These pass to the right and to the left to the extremes of the superior liver surface, turn back on themselves (creating the **triangular ligaments**), and turn posteriorly to form the posterior layers of the coronary ligaments. In this manner, an area devoid of visceral peritoneum is created, the **bare area of the liver.** The posterior layers of the coronary ligaments converge to form the **lesser omentum,** which passes from the visceral surface of the liver to the lesser curvature of the stomach (**hepatogastric ligament**) and the first part of the duodenum (**hepatoduodenal ligament**).

The **liver** is divided anatomically into **four** lobes **by external landmarks** and is delineated on the visceral surface by fissures and fossae, which form the letter H (see Figure 23-1). The **right side of the H** is formed by **fossae for the gallbladder and the IVC,** and the **right lobe lies to the right** of these structures. The **left side of the H** is formed by the **fissure for the round ligament and the ligamentum venosum** (adult remnant of the ductus venosus); the **left lobe is to the left** of this fissure. The **crossbar of the H** is the **porta hepatis,** through which **the hepatic artery, portal vein, and nerves** enter the liver and the **bile ducts and lymphatics exit.**

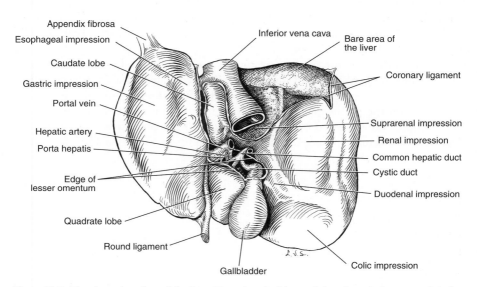

Figure 23-1. The visceral surface of the liver. (*Reproduced, with permission, from Lindner HH. Clinical Anatomy. East Norwalk, CT: Appleton & Lange, 1989:399.*)

The crossbar subdivides the central portion into **quadrate** and **caudate lobes.** Functionally, the right portal lobe lies to the right of the fossae of the gallbladder, IVC, and a portion of the caudate lobe. The left portal lobe is the left anatomical lobe, quadrate lobe, and the remainder of the caudate lobe. The portal lobes are supplied by lobar branches of the hepatic artery, portal vein, and bile ducts. Although lacking external landmarks, the portal lobes are further divided functionally into hepatic segments.

The liver receives a **dual blood supply;** approximately **30 percent of the blood entering the organ is from the hepatic artery, and 70 percent is from the portal vein.** The **proper hepatic artery** is a branch of the common hepatic artery, one of the three major branches of the celiac artery. As it approaches the liver, it divides into right and left hepatic branches that enter the liver and divide into lobar, segmental, and smaller branches. Eventually, blood reaches the arterioles in the portal areas at the periphery of the hepatic lobules and, after providing oxygen and nutrients to the parenchyma, drains into the hepatic sinusoids. The majority of blood entering the liver is venous blood rich in nutrients and molecules absorbed by the gastrointestinal organs. Intrahepatic branches of the portal vein follow the arteries to the portal areas, where portal venules empty into the sinusoids from which molecules are extracted and added. Sinusoidal blood flows to the central vein of each lobule from which increasingly large veins are formed until typically three hepatic veins exit the liver to join the IVC (Figure 23-2).

The **portal venous system** arises from the capillary beds within the abdominal organs supplied by the **celiac artery, superior mesenteric artery (SMA), and inferior mesenteric artery (IMA),** and blood will flow to and through the liver for metabolism of its contained molecules. Veins from these organs for the most part accompany arteries of the same name. The **portal vein itself is formed by the union of the splenic vein and SMV posterior to the neck of the pancreas.** This short, wide vein ascends within the hepatoduodenal ligament, posterior to the bile duct and hepatic artery, and enters the liver through the porta hepatis. Typically, the IMV drains its blood into the splenic vein.

Portacaval (systemic) venous anastomoses occur at sites where blood may ultimately **drain into the portal system and/or the vena caval system.** If venous flow through the portal system is prevented by liver disease, for example, the absence of valves within the portal system veins allows reverse flow. This dilates the smaller veins, and blood is drained by veins emptying into the vena cavae. This occurs at several sites and may produce clinical signs or symptoms (Table 23-1).

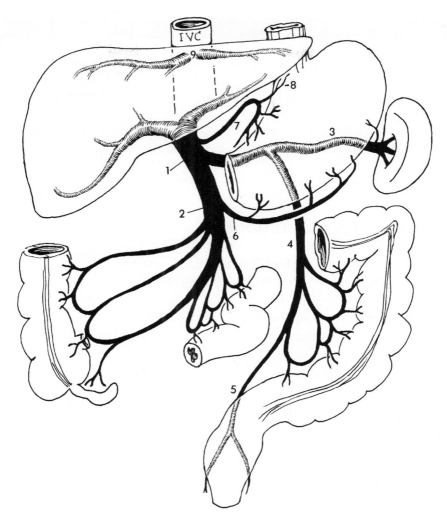

Figure 23-2. The portal system: 1 = portal vein, 2 = superior mesenteric vein, 3 = splenic vein, 4 = inferior mesenteric vein, 5 = superior rectal vein, 6 = right gastroepiploic vein, 7 = left gastric vein, 8 = esophageal vein, 9 = hepatic veins. (*Reproduced, with permission, from the University of Texas Health Science Center Houston Medical School.*)

Table 23-1 • SITES OF PORTAL-CAVAL VENOUS ANASTOMOSES			
	Portal Venous Drainage	Vena Cava Venous Drainage	Sign/Symptom
Esophagus	Left gastric vein	Hemiazygous vein	Esophageal varices, bleeding
Rectum	Superior rectal vein	Inferior rectal vein	Hemorrhoids
Anterior abdominal wall	Paraumbilical vein	Intercostal vein	Caput medusa
Retroperitoneal	Duodenal, pancreatic, right and left colic veins	Lumbar vein	Intestinal bleeding

COMPREHENSION QUESTIONS

23.1   You are examining the liver during a surgical procedure. The gallbladder will be found in its fossa between which two anatomical lobes?

   A.   Quadrate and left lobes

   B.   Quadrate and caudate lobes

   C.   Right and quadrate lobes

   D.   Caudate and right lobes

   E.   Caudate and left lobes

23.2   If you ligated the right hepatic artery, the arterial supply to which of the following portions of the liver would *remain* intact?

   A.   Right lobe only

   B.   Right and quadrate lobes

   C.   Left lobe only

   D.   Left and quadrate lobes only

   E.   Left, quadrate, and a portion of the caudate lobe

23.3   Your patient who had cirrhosis has symptoms of esophageal varices. This is due to dilatation of the anastomosis between which of the following pairs of veins?

   A.   Left gastric and azygous veins

   B.   Right gastric and azygous veins

   C.   Right gastric and hemiazygous veins

   D.   Left gastric and hemiazygous veins

   E.   Azygous and hemiazygous veins

ANSWERS

23.1 **C.** The gallbladder is located between the right lobe and the quadrate lobe.

23.2 **E.** The left hepatic artery supplies the left and quadrate lobes and a portion of the caudate lobe.

23.3 **D.** Esophageal veins drain to the left gastric and hemiazygous veins.

ANATOMY PEARLS

▶ The left anatomical lobe, quadrate lobe, and a portion of the caudate lobe constitute the left portal lobe.

▶ Hemorrhage from the liver can be controlled by clamping the hepatoduodenal ligament (Pringle maneuver), which contains the hepatic artery and portal vein.

▶ The portal vein drains blood from organs supplied by the celiac artery, superior mesenteric artery (SMA), and inferior mesenteric artery (IMA).

▶ Esophageal varices with bleeding is the most clinically significant symptom of portal hypertension.

REFERENCES

Gilroy AM, MacPherson BR, Ross LM. *Atlas of Anatomy,* 2nd ed. New York, NY: Thieme Medical Publishers; 2012:164−167.

Moore KL, Dalley AF, Agur AMR. *Clinically Oriented Anatomy,* 7th ed. Baltimore, MD: Lippincott Williams & Wilkins; 2014:268−277, 285−286.

Netter FH. *Atlas of Human Anatomy,* 6th ed. Philadelphia, PA: Saunders; 2014: plates 277−279, 291−292.

A 42-year-old male executive complains of abdominal pain that began about 6 months previously; is constant in nature, especially after meals; and is located in the upper midabdomen superior to the umbilicus. He also reports some heartburn that occurred during the previous year. He has been under significant job-related stress and has been self-medicating himself with over-the-counter antacids, with some relief. He states that his stools have changed in color over the previous 2 months and now are intermittently dark and tarry in consistency. The physician tests the patient's stool and finds occult blood.

▶ What is the most likely diagnosis?
▶ What organs are likely to be affected?

ANSWER TO CASE 24:
Peptic Ulcer Disease

Summary: A 42-year-old stressed-out male executive has a 6-month history of constant upper abdominal pain and heartburn for the past year that was relieved by over-the-counter antacids. His stools have become dark and tarry, which upon examination contain occult blood.

- **Most likely diagnosis:** Peptic ulcer disease

- **Organs likely affected:** Stomach or duodenum

CLINICAL CORRELATION

This patient has a history typical for peptic ulcer disease, that is, constant mid-epigastric pain after meals. The patient also has symptoms consistent with gastro-esophageal reflux disease. The dark and tarry stools reflect blood in the stools; that is, hemoglobin has been converted to melena. This is suggestive of an upper gastro-intestinal bleeding disorder. The next step would be an upper endoscopy to visualize the suspected ulcer. If the stomach is the site, a biopsy is usually performed to assess concurrent malignancy. Treatment includes a histamine-blocking agent, proton pump inhibitor, and antibiotic therapy. The bacterium *Helicobacter pylori* has been implicated in most cases of peptic ulcer disease. If an ulcer occurs in the duodenum, the posterior wall of the ampulla of the duodenum (duodenal cap) is the usual site. The gastroduodenal artery lies posterior to the duodenum at this point and is at risk in the event of ulcer perforation.

APPROACH TO:
The Stomach

OBJECTIVES

1. Be able to describe the anatomy of the stomach

2. Be able to describe the anatomy of the celiac artery (trunk)

DEFINITIONS

GASTROESOPHAGEAL REFLUX DISEASE: Condition in which gastric contents are regurgitated into the esophagus

PEPTIC ULCER DISEASE: A lesion of the gastric or duodenal mucosa with inflammation

HELICOBACTER PYLORI: Bacterium found in the mucosa of humans and associated with peptic ulcer disease

ENDOSCOPY: Procedure by which the interiors of hollow organs are examined with a flexible instrument called an **endoscope**

DISCUSSION

The **stomach,** the **first major gastrointestinal organ in which digestion occurs, produces digestive enzymes and hydrochloric acid (HCl).** This continuation of the esophagus is a large, intraperitoneal, saccular organ that is suspended by the mesentery-like greater and lesser omenta. The stomach is divided anatomically into a **cardia, fundus, body, and pylorus (pyloric antrum and canal with sphincter)** and has **greater and lesser curvatures.** The **greater omentum** attaches to the greater curvature and drapes inferiorly to form a double-layer apron anterior to the abdominal cavity contents. It fuses superiorly with the transverse mesocolon. The greater omentum is subdivided into gastrocolic, gastrosplenic, splenorenal, and gastrophrenic ligament portions. The **lesser omentum** is attached to the lesser curvature and first part of the duodenum and extends to the visceral surface of the liver. With the stomach, it forms the anterior boundary of the omental bursa (lesser sac). The lesser omentum is divided into hepatogastric and hepatoduodenal ligaments; the latter form the anterior boundary of the **epiploic foramen (of Winslow;** see Figure 24-1).

The stomach is richly supplied by **five sets of arteries,** all of which are branches of the **celiac artery (trunk).** The celiac artery arises from the abdominal aorta opposite the upper portion of L1. This very short artery quickly divides into three branches.

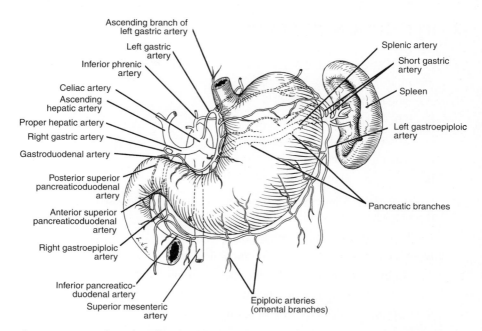

Figure 24-1. Arterial supply to the stomach. (*Reproduced, with permission, from Lindner HH. Clinical Anatomy. East Norwalk, CT: Appleton & Lange, 1989:334.*)

The smallest is the **left gastric artery,** which ascends toward the gastroesophageal junction at the lesser curvature. After sending small branches to the esophagus, it curves inferiorly within the lesser omentum, parallel to the lesser curvature, to which numerous **gastric branches** are provided. The **splenic artery is a large, tortuous branch** of the celiac that passes to the left, along the superior margin of the pancreas, to reach the spleen. It sends several branches to the pancreas and, as it approaches the spleen, gives off two sets of arteries to the stomach. Passing superiorly, **four to five small short gastric arteries** ascend within the gastrosplenic ligament to supply the fundus. Also near the spleen, the **left gastroomental (epiploic) artery** arises from the splenic artery and passes inferiorly within the gastrosplenic and gastrocolic ligaments. It courses parallel to the greater curvature, to which it sends numerous branches. The last branch of the celiac is the **common hepatic artery,** which passes to the right to enter the hepatoduodenal ligament. The common hepatic divides into two branches. The proper hepatic artery ascends toward the liver within the hepatoduodenal ligament to supply the liver and gallbladder. The **right gastric artery** typically arises from the proper hepatic, descends to the gastroduodenal junction, curves superiorly and parallel to the lesser curvature, sends gastric branches to the stomach, and **anastomoses with the left gastric artery.** The other branch of the common hepatic artery is the **gastroduodenal artery,** which descends posterior to the first part of the duodenum and then divides into the **pancreaticoduodenal** and **right gastroomental arteries.** This latter vessel lies within the gastrocolic ligament and courses to the left, parallel to the greater curvature, to which gastric branches are sent. It anastomoses with the left gastroomental artery along the greater curvature.

COMPREHENSION QUESTIONS

24.1 Gastric contents exiting a posterior perforation of the stomach wall will accumulate in which of the following?

  A. The left paracolic gutter

  B. The left paravertebral gutter

  C. The right paravertebral gutter

  D. The omental bursa

  E. The hepatorenal recess

24.2 Ligation of the common hepatic artery will eliminate the gastric blood supply through which of the following arteries?

  A. Left gastric and short gastric arteries

  B. Short gastric and right gastroomental arteries

  C. Right gastroomental and right gastric arteries

  D. Right gastric and left gastric arteries

  E. Left gastric and left gastroomental arteries

24.3 A surgical incision through the fundus of the stomach would require you to clamp which of the following?

A. Right gastric artery

B. Left gastric artery

C. Right gastroomental artery

D. Left gastroomental artery

E. Short gastric arteries

24.4 A 45-year-old woman is brought to the emergency department (ED) with a 4-h history of vomiting dark brown emesis that has a granular component. Examination shows tenderness of the midabdomen. Upper endoscopy shows a bleeding ulcer of the duodenal bulb on the posterior aspect. Which artery has the ulcer most likely have affected?

A. Splenic

B. Right gastroepiploic

C. Left gastric

D. Gastroduodenal

E. Celiac trunk

ANSWERS

24.1 **D.** The omental bursa lies immediately posteriorly to the stomach.

24.2 **C.** Blood flow through the right gastric and right gastroomental arteries would be lost with ligation of the common hepatic artery.

24.3 **E.** The short gastric arteries supply the fundus of the stomach.

24.4 **D.** The gastroduodenal artery arises from the common hepatic artery and supplies the proximal duodenum. Although duodenal ulcers are typically anterior (which can lead to perforation), deep duodenal ulcers on the posterior aspect can erode into the gastroduodenal artery and lead to significant bleeding.

ANATOMY PEARLS

▶ The relatively fixed points of the stomach are the gastroesophageal junction and the pylorus, which lie at vertebral levels T11 and L1, respectively.

▶ The stomach is supplied by all three branches of the celiac artery.

▶ The short gastric and left gastroomental arteries lie within the gastrosplenic ligament and are at risk in a splenectomy.

REFERENCES

Gilroy AM, MacPherson BR, Ross LM. *Atlas of Anatomy*, 2nd ed. New York, NY: Thieme Medical Publishers; 2012:156–157.

Moore KL, Dalley AF, Agur AMR. *Clinically Oriented Anatomy*, 7th ed. Baltimore, MD: Lippincott Williams & Wilkins; 2014:230–237, 256.

Netter FH. *Atlas of Human Anatomy*, 6th ed. Philadelphia, PA: Saunders; 2014: plates 266–268, 283.

A 55-year-old male is admitted to the hospital for a suspected kidney infection. He is placed on intravenous antibiotic therapy but continues to have a temperature of 103°F after 3 days of therapy. The urine culture grows *Escherichia coli*, which is sensitive to the antibiotics being used. On examination, he appears ill and has marked left flank tenderness. Ultrasound depicts an abnormal collection of fluid around the left kidney.

▶ What is the most likely diagnosis?
▶ What anatomical structure is involved?

ANSWER TO CASE 25:
Perinephric Abscess

Summary: A 55-year-old male continues with high fever and flank pain despite 3 days of broad-spectrum intravenous antibiotic therapy. The urine isolate of *E. coli* demonstrates in vitro sensitivity to the antibiotics used. Renal ultrasound shows fluid around the left kidney.

- **Most likely diagnosis:** Perinephric abscess
- **Anatomical structure involved:** Kidney and anatomically related structures

CLINICAL CORRELATION

This 55-year-old male who is suspected of having pyelonephritis is not improving despite appropriate antibiotic therapy. Pyelonephritis is an infection of the kidney parenchyma usually caused by an ascending infection of bacteria that advances from the urethra to the bladder, to the ureters, and then to the kidney. Kidney infection usually manifests as fever, flank tenderness, white cells in the urine, and serum leukocytosis. After 48 to 72 h, one would expect decreases in fever and flank tenderness. *E. coli* is isolated, which is the bacterium that most often causes urinary tract infections. The ultrasound examination is performed to rule out complications of pyelonephritis. The two most common complications would be a nephrolithiasis or ureterolithiasis (kidney stone) and perinephric abscess. Intervention is required before improvement is seen. The abscess must be drained, usually by placement of a percutaneous catheter under radiologic guidance.

APPROACH TO:
The Kidneys

OBJECTIVES

1. Be able to describe the anatomy of the kidneys and their fascial coverings and blood supply
2. Be aware of the structures next to the kidneys and their relations

DEFINITIONS

PERINEPHRIC ABSCESS: Collection of pus in the tissues surrounding the kidney

PYELONEPHRITIS: Usually a bacterial inflammation of the renal tissue, the calyces, or renal pelvis

NEPHROLITHIASIS: Presence of renal calculi or stones

DISCUSSION

The **kidneys are paired retroperitoneal organs** that are located in the **paravertebral gutters.** The **left kidney lies slightly higher than the right,** its hilum is at the **level of L1,** and its superior and inferior poles are at the 11th rib and L3, respectively. The hilum of the **right kidney lies at the level of the disk between L1 and L2,** and its inferior pole is nearly 1 to 2 cm superior to the iliac crest. Each kidney is an encapsulated solid organ, with an outer cortex and an inner medulla, with the latter arranged in renal pyramids. The hilum of each kidney leads to a space, the renal sinus, which contains fat, branches of the renal vessels, and the urine-collecting structures (minor and major calyces and renal pelvis). Within the sinus, the apex of the 6 to 12 renal pyramids is cupped by a minor calyx, which collects the urine produced. Typically, two to three minor calyces unite to form a major calyx, and two to three major calyces form the renal pelvis. The renal pelvis is continuous with the ureter at the inferior margin of the hilum (see Case 32 for the anatomy of the ureter).

Four muscles are related to each kidney posteriorly: the diaphragm superiorly and the transversus abdominis, quadratus lumborum, and psoas muscles inferiorly, from lateral to medial. The **suprarenal glands and colon contact both kidneys anteriorly.** The **duodenum and liver also contact the right kidney anteriorly,** and the **stomach, pancreas, and spleen are related to the anterior left kidney.**

Each **kidney and suprarenal gland is encased in a renal (Gerota) fascia** (Figure 25-1), which helps to maintain the position of the kidney. The renal fascia fuses with the fascia of the psoas muscle posteriorly and with the adventitia of the renal vessels anteriorly. Within the renal fascia is an accumulation of fat known as **perirenal fat,** which is continuous with the fat within the renal sinus. **Pararenal fat** surrounds each kidney external to the renal fascia. Pararenal fat is thick posterior to the kidney, but it is thin anteriorly between the renal fascia and parietal peritoneum.

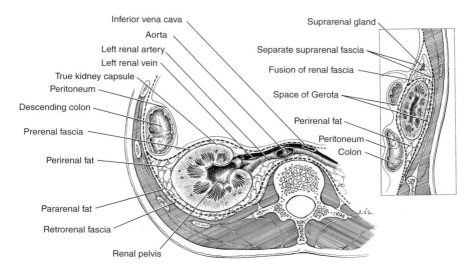

Figure 25-1. The left kidney and surrounding fascia. (*Reproduced, with permission, from Lindner HH. Clinical Anatomy. East Norwalk, CT: Appleton & Lange, 1989:444.*)

Each kidney is supplied by a renal artery that arises from the aorta near vertebral level L2. As each artery nears the renal pelvis, it typically divides into five segmental arteries that enter the hilum to supply segments of renal tissue. The **right renal artery is the longer artery**, and **both renal arteries lie posterior to the renal veins** when entering the hilum. The renal veins exit the hilum anterior to the arteries, and the left vein is longer and crosses the midline. **Both renal veins drain into the IVC.** The **left renal vein is unique in that the inferior phrenic, suprarenal, and gonadal veins drain into it** (the IVC receives these veins on the right side).

COMPREHENSION QUESTIONS

25.1 During the removal of a patient's kidney, you would observe which of the following as being most anterior within the renal sinus?

A. Renal arteries

B. Renal vein

C. Major calyx

D. Minor calyx

E. Renal pelvis

25.2 You wish to examine the hilum of the right kidney during surgery. Which of the following structures must be elevated and reflected to do so?

A. Stomach

B. Suprarenal gland

C. Ascending colon

D. Duodenum

E. Liver

25.3 To elevate the kidney within the renal fascia and the perirenal fat, the renal fascia must be reflected or incised from the fascia of which of the following muscles?

A. Diaphragm

B. Psoas muscle

C. Quadratus lumborum muscle

D. Transversus abdominis muscle

E. Iliacus muscle

ANSWERS

25.1 **B.** The renal veins lie most anterior within the renal sinus.

25.2 **D.** The duodenum lies immediately anteriorly to the hilum of the right kidney.

25.3 **B.** The renal fascia is fused posteriorly to the fascia of the psoas muscle.

ANATOMY PEARL

▶ The hilum of the left kidney lies at the level of L1.

REFERENCES

Gilroy AM, MacPherson BR, Ross LM. *Atlas of Anatomy,* 2nd ed. New York, NY: Thieme Medical Publishers; 2012:172–175, 236.

Moore KL, Dalley AF, Agur AMR. *Clinically Oriented Anatomy,* 7th ed. Baltimore, MD: Lippincott Williams & Wilkins; 2014: 290–292, 298.

Netter FH. *Atlas of Human Anatomy,* 6th ed. Philadelphia, PA: Saunders; 2014: plates 308–310, 315.

An 18-year-old female presents with increasing hair growth on her face and chest, deepening of her voice, and acne over the past year. She has no history of other medical problems. On examination, she has acne, abnormal male pattern balding, and enlargement of her clitoris. The pelvic examination is normal including the ovaries. Blood tests show normal serum testosterone levels but a markedly elevated level of dihydroepiandrostenedione sulfate, an adrenal androgen.

▶ What is the most likely diagnosis?

ANSWER TO CASE 26:
Suprarenal Gland Tumor

Summary: An 18-year-old female is seen for increasing hirsutism, deepening voice, and acne over the past year. She has no other medical problems. On examination she displays acne, hirsutism, temporal balding, and clitoromegaly. The pelvic examination, including the ovaries, is normal. The testosterone level is normal, and the level of dihydroepiandrostenedione sulfate is markedly elevated.

- **Most likely diagnosis:** Suprarenal (adrenal) gland tumor

CLINICAL CORRELATION

This young female has more than hirsutism, which is increased hair growth. She also has virilism, or the effects of androgens on the skin, voice, and clitoris. The hyperandrogenism seems to be of acute onset, which is consistent with an androgen-secreting tumor. The two possibilities include an ovarian tumor, usually Sertoli-Leydig cell tumor, or an adrenal tumor. Because the pelvic examination and testosterone levels are normal, an ovarian etiology is less likely. Moreover, the high level of dihydroepiandrostenedione sulfate almost establishes the suprarenal (adrenal) gland as the cause. The next step would be a CT or MRI scan of the suprarenal glands to determine the exact location of the tumor. Usually surgery is indicated. Another cause of hirsutism is polycystic ovarian syndrome, which includes hirsutism, obesity, anovulation, and irregular menses. Cushing syndrome or disease presents strong cortisol effects, such as buffalo hump, abdominal striae, easy bruising, and central obesity.

APPROACH TO:
The Suprarenal Glands

OBJECTIVES

1. Be able to describe the anatomy of the suprarenal glands
2. Be able to describe the general pattern of lymphatic drainage of the abdomen

DEFINITIONS

VIRILISM: Presence of mature male characteristics in a female or a prepubescent male

CLITOROMEGALY: Enlargement of the clitoris

DIHYDROEPIANDROSTENEDIONE (DHEA): Male steroid hormone secreted by the testis, ovary, or adrenal cortex

DISCUSSION

The **paired suprarenal glands are retroperitoneal endocrine glands** composed of an **outer cortex that secretes corticosteroid and androgen steroid hormones** and an **inner medulla (derived from neural crest cells) that secretes the catecholamines, epinephrine and norepinephrine.** Each gland **sits on the superior pole of each kidney, enclosed within the renal fascia,** and, hence, embedded in the perirenal fat. The **right suprarenal gland** is somewhat triangular and is closely related to the IVC, liver, and diaphragm. The **left gland is shaped like a comma** and related to the spleen, pancreas, stomach, and diaphragm. The suprarenal glands receive their blood supply from multiple small branches that arise from the inferior phrenic, aorta, and renal arteries. Each gland is drained by a single suprarenal vein that terminates in the IVC on the right and the renal vein on the left.

The **lymphatic drainage of the abdomen** is diagrammatically summarized in Figure 26-1. In general, the lymphatic drainage of abdominal organs reversely follows their arterial blood supply. Thus the **lymph drainage from organs supplied by the SMA will be to the superior mesenteric nodes** by way of vessels and other node groups located along the branches of the SMA. If a "final common pathway" **for lymph drainage** in the abdomen could be named, it would be the **lumbar (aortic) lymph nodes,** and lymph from these nodes drains to the **cisterna chyli and thoracic duct.** Lymph from the **suprarenal glands drains into the lumbar lymph nodes.** Figure 26-1 shows that the lymphatics from the gonads also drain to the upper lumbar nodes as the gonadal vessels arise in the upper abdomen (reflecting the site of their embryologic origin). Note also that the **pectinate line in the anal canal is a watershed with regard to lymphatic drainage.** The lymph from the anal canal and rectum **superior to this line drains to iliac nodes; inferior to this line, lymph drains to inguinal nodes.**

Lymphatic drainage of the abdomen

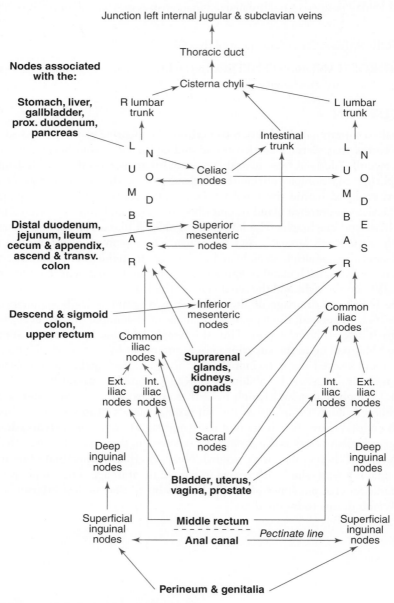

Figure 26-1. Lymph node drainage of the abdomen.

COMPREHENSION QUESTIONS

26.1 As a surgeon about to remove the right adrenal gland, you examine the blood supply of the right adrenal gland and observe which of the following?

  A. It receives its arterial blood supply from the aorta only

  B. Its central vein drains into the IVC

  C. Its central vein drains into the left renal vein

  D. It is in contact with the head of the pancreas

  E. It lies external to the renal fascia

26.2 After removal of a large portion of the stomach in a patient who has cancer, you are now examining the lymph nodes that receive lymph from the stomach. Which of the following structures receives lymph directly from the stomach?

  A. Cisterna chyli

  B. Aorticorenal nodes

  C. Celiac nodes

  D. Superior mesenteric nodes

  E. Inferior mesenteric nodes

26.3 In a patient who has testicular cancer that has metastasized (spread) to the lymph nodes, which of the following would you expect to be involved first?

  A. Lumbar (aortic) nodes

  B. Aorticorenal nodes

  C. Inferior mesenteric nodes

  D. Common iliac nodes

  E. Internal iliac nodes

ANSWERS

26.1 **B.** The central vein of the right suprarenal gland drains into the IVC, whereas that of the right gland drains into the left renal vein.

26.2 **C.** The lymph of nodes located along the several arteries that supply the stomach will drain to the celiac nodes. Remember that the arteries that supply the stomach are all branches of the celiac artery.

26.3 **A.** Tumor cells from either gonad that metastasize through the lymphatics will metastasize to the lumbar (aortic) lymph group. Remember that the origin of the gonadal arteries is the abdominal aorta.

ANATOMY PEARLS

▶ The right suprarenal gland is closely related to the IVC, into which its vein will open.

▶ The left suprarenal vein drains into the left renal vein.

▶ Multiple arteries to the suprarenal glands arise from the inferior phrenic artery, aorta, and renal artery.

▶ Lymph from the gonads drains to upper lumbar nodes.

▶ Lymph above the pectinate line of the anal canal drains to the iliac nodes, whereas lymph below the pectinate line drains to inguinal nodes.

REFERENCES

Gilroy AM, MacPherson BR, Ross LM. *Atlas of Anatomy,* 2nd ed. New York, NY: Thieme Medical Publishers; 2012:172–175, 236.

Moore KL, Dalley AF, Agur AMR. *Clinically Oriented Anatomy,* 7th ed. Baltimore, MD: Lippincott Williams & Wilkins; 2014:294–297.

Netter FH. *Atlas of Human Anatomy,* 6th ed. Philadelphia, PA: Saunders; 2014: plates 261, 308, 310–311, 319.

A 27-year-old female notes a tender lump in her groin area that appeared approximately 3 weeks ago. She relates that she had a similar mass about 1 year ago that required minor surgery. On physical examination, she is afebrile, and inspection of the perineum shows a 3 × 2-cm fluctuant mass at the five-o'clock position of the vestibule. It is mildly tender, red, and slightly warm to the touch.

▶ What is the most likely diagnosis?
▶ What structures are causing the groin enlargement?

ANSWER TO CASE 27:

Greater Vestibular (Bartholin) Gland Abscess

Summary: A 27-year-old female notes a tender lump in the groin that appeared 3 weeks ago. She had surgery for a similar mass a year ago. The patient is afebrile and has a 3 × 2-cm fluctuant, inflamed mass at the five-o'clock position of the vestibule.

- **Most likely diagnosis:** Greater vestibular (Bartholin) gland abscess
- **Cause of groin lump:** Inguinal lymph nodes

CLINICAL CORRELATION

This young woman notes the appearance of an inflamed, perineal, or vulvar mass in the posterolateral region of the vestibule. She apparently had a similar lesion 1 year previously. These findings are very consistent with a greater vestibular (Bartholin) gland infection. The greater vestibular glands are located at the five- and seven-o'clock positions of the vulva. If the ducts of the glands become obstructed, the glands may enlarge and become infected, usually with multiple organisms other than those responsible for sexually transmitted diseases. Lymphatic drainage of the vulva is first to the inguinal lymph nodes. Treatment for this patient is to create a fistulous tract to decrease the incidence of recurrence; the two most common methods are incision and drainage with a catheter left in place for several weeks and marsupialization of the cyst wall, which is suturing the inner lining of the cyst wall to the epithelium around the periphery of the cyst. Biopsy is typically not required in a young patient, but for vulvar masses or abnormalities in women older than 40 years it is required to rule out malignancy.

APPROACH TO:

The Vulva

OBJECTIVES

1. Be able to define the boundaries of the perineum
2. Be able to describe the urogenital triangle
3. Be able to describe the lymphatic drainage of the perineum

DEFINITIONS

VULVA: The region of the external genitalia of the female

GREATER VESTIBULAR GLANDS: Bartholin glands; two small reddish bodies on the posterolateral aspects of the vestibule

MARSUPIALIZATION: Surgical procedure in which the inner lining of the cyst wall is sutured to the epithelium around the periphery of the cyst to promote cyst drainage

DISCUSSION

The **perineum** is defined as the region of the trunk, between the thighs and buttocks, inferior to the pelvic diaphragm. It is bounded bilaterally by the **pubic symphysis** (anterior), **ischiopubic ramus** (anterolateral), **ischial tuberosity** (lateral), **sacrotuberous ligament** (posterolateral), and the **coccyx** (posterior). A line between the ischial tuberosities divides the perineum into **anterior and posterior urogenital** and **anal triangles,** respectively. Deep to the skin is the fatty layer of superficial fascia, a continuation of a similar layer in the abdomen (**Camper fascia**). In the abdomen, deep to the fatty layer is the membranous layer of superficial fascia (Scarpa fascia) that continues into the perineum, where it is called **Colles fascia.** In the perineum, Colles fascia is attached laterally to the fascia lata of the thigh and to the posterior border of the perineal membrane and the perineal body. The perineal membrane is a thin but strong fascial sheet attached to the ischiopubic rami, thus stretching across the urogenital triangle. The potential space between the deep layer of the **superficial (Colles) fascia** and the perineal membrane is the superficial perineal pouch (space). Attached to the superior surface of the perineal membrane are the **deep transverse perineal and sphincter urethrae muscles** within the **deep perineal pouch** (space). The perineal body attaches to the posterior edge of the membrane at its midpoint (Figures 27-1 and 27-2).

Superficial to the perineal membrane, the **pudendum or vulva (external genitalia)** includes the **mons pubis** and **labia majora, labia minora, vaginal vestibule, bulbs of the vestibule, greater vestibular (Bartholin) glands, clitoris,** and the associated **ischiocavernosus and bulbospongiosus muscles.** The mons pubis is a rounded, hair-covered elevation anterior to the pubic symphysis formed by a mass of the fatty layer of the superficial fascia. Fat-filled posterior extensions of the mons form the hair-covered labia majora, which are united by anterior and posterior commissures.

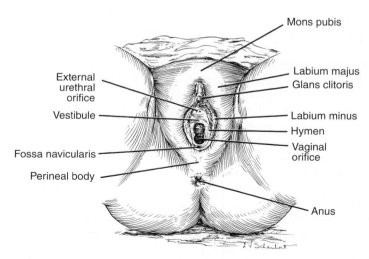

Figure 27-1. External female genitalia. (*Reproduced, with permission, from Decherney AH, Nathan L. Current Obstetric and Gynecologic Diagnosis and Treatment, 9th ed. New York: McGraw-Hill, 2003:17.*)

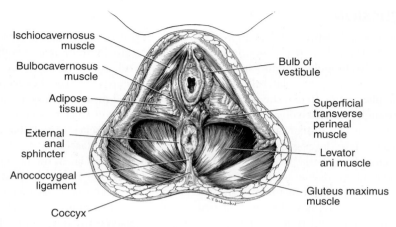

Figure 27-2. Interior view of female pelvic musculature. (*Reproduced, with permission, from Decherney AH, Nathan L. Current Obstetric and Gynecologic Diagnosis and Treatment, 9th ed. New York: McGraw-Hill, 2003:22.*)

The space between the two labia is the **pudendal cleft.** Medial to each labia majora are the thin, fat-free, hairless **labia minora** that are filled with erectile tissue and surround the vestibule of the vagina, which contains the **urethral and vaginal orifices.** The labia minora are united posteriorly by **the frenulum of the labia minora or fourchette.** Anteriorly, the two labia minora are united by extensions that pass anterior and posterior to the glans of the clitoris as the prepuce and frenulum of the clitoris, respectively. The clitoris is composed **of paired cylinders of erectile tissue or corpora cavernosa attached to the ischiopubic rami as the two crura** and are surrounded by the ischiocavernosus muscles. The **corpora cavernosa** converge toward the pubic symphysis to form the body, which is sharply flexed inferiorly and terminates as the glans anterior to the urethral orifice. Superior (deep) to the labia majora and minora, at the margins of the vestibule, are the **paired bulbs of the vestibule.** At the posterior ends of the bulbs and partially embedded in them are the paired **greater vestibular (Bartholin) glands.** The bulbs and glands are covered by the bulbospongiosus muscles. A superficial transverse perineal muscle lies along the posterior edge of the perineal membrane and attaches laterally to the ischial tuberosity and medially to the perineal body. The components of the clitoris, bulb of the vestibule, greater vestibular gland, and the bulbospongiosus and ischiocavernosus muscles are encased in the deep perineal or investing (Gallaudet) fascia. The bulbospongiosus, superficial, and deep transverse perineal, and external anal sphincter muscles attach to the perineal body.

The **lymphatic drainage** of the perineum is primarily to the **superficial inguinal lymph nodes,** which lie inferior and parallel to the inguinal ligament. Efferent vessels from this group drain lymph to the **external iliac nodes,** but some lymph does drain to the **deep inguinal nodes,** which then drain to the external iliac nodes. Small amounts of lymph from deep perineal structures drain to the internal iliac nodes.

COMPREHENSION QUESTIONS

27.1 A 34-year-old woman who has diabetes develops a boil on the right labia majora. Which of the following lymph nodes is most likely to be enlarged in response to the infection?

A. Internal iliac

B. External iliac

C. Superficial inguinal

D. Obturator

27.2 Which of the following structures divides the perineum into the genitourinary and anal triangles?

A. Levator ani muscles

B. Superficial transverse perineal muscle

C. Line from the ischial tuberosities

D. Anal verge

27.3 A 24-year-old woman is undergoing a vaginal delivery. A midline episiotomy is performed that incises into the perineal body. Which of the following muscles is most likely to be cut during this process?

A. Superficial transverse perineal muscle

B. Levator ani muscle

C. Puborectalis muscle

D. Pubococcygeus muscle

ANSWERS

27.1 **C.** The primary drainage of the vulva is the superficial inguinal nodes.

27.2 **C.** A line between the two ischial tuberosities divides the perineum into the genitourinary triangle (anteriorly) and the anal triangle (posteriorly).

27.3 **A.** The muscles that attach to the perineal body are the bulbospongiosus, superficial, and deep transverse perineal muscles, and the external anal sphincter.

ANATOMY PEARLS

▶ The clitoral structures, vestibular bulb, greater vestibular gland, and associated muscles are located in the superficial perineal pouch (space).

▶ The bulbospongiosus, ischiocavernosus, superficial, and deep transverse perineal, and sphincter urethral muscles are innervated by the pudendal nerve.

▶ The primary lymphatic drainage of the perineum is to the superficial inguinal lymph nodes.

REFERENCES

Gilroy AM, MacPherson BR, Ross LM. *Atlas of Anatomy,* 2nd ed. New York, NY: Thieme Medical Publishers; 2012:248, 267.

Moore KL, Dalley AF, Agur AMR. *Clinically Oriented Anatomy,* 7th ed. Baltimore, MD: Lippincott Williams & Wilkins; 2014:402–406, 428–431, 433.

Netter FH. *Atlas of Human Anatomy,* 6th ed. Philadelphia, PA: Saunders; 2014: plates 354, 356–357, 382.

A 20-year-old male reports that he has had a nontender, heavy sensation in his scrotal area for 2 months. He jogs several miles every day but denies lifting heavy objects. He does not recall trauma to the area, has no urinary complaints, does not smoke, and otherwise appears healthy. His blood pressure is 110/70 mmHg, his heart rate is 80 beats/min, and he is afebrile. Heart and lungs examinations are normal. His back and abdomen are nontender, and no abdominal masses are detected. External genitalia examination reveals a 2-cm nontender mass in the right testicle that shows no light penetration with transillumination. The rectal examination is unremarkable.

▶ What is the most likely diagnosis?

ANSWER TO CASE 28:

Testicular Cancer

Summary: A 20-year-old male is noted to have a nontender heavy sensation in the scrotal area of 2 months' duration. He jogs several miles each day and denies lifting heavy objects, scrotal trauma, and urinary problems. A 2-cm, nontender, nontransilluminating mass is noted in the right testicle. The rectal examination is unremarkable.

- **Most likely diagnosis:** Testicular cancer

CLINICAL CORRELATION

Testicular carcinoma affects young men, usually between ages 15 and 40 years, and the presence of a painless scrotal mass is the most common presentation. A history of trivial scrotal trauma is not uncommon, which often brings the scrotal mass to the patient's attention. Testicular carcinoma should be ruled out before other conditions are considered, such as varicocele, spermatocele, hydrocele, epididymitis, or testicular torsion. Regular scrotal examination is advocated but rarely performed, and personal embarrassment often delays medical consultation.

APPROACH TO:

Male Genitalia

OBJECTIVES

1. Be able to describe the anatomy of the external male genitalia
2. Be able to draw the blood supply and lymphatic drainage of the testicles

DEFINITIONS

TRANSILLUMINATION: Passage of light through a specific tissue during examination with the object between the light source and the examiner

CIRCUMCISION: Removal of all or part of the prepuce or foreskin

HYDROCELE: Collection of fluid in the tunica vaginalis of the testicle or along the spermatic cord

DISCUSSION

The **male external genitalia** consist of the **penis** and the **scrotum**, which contains the **testes**, the male gonad. All of these structures lie within the boundaries of the **urogenital triangle** of the **perineum.** The relations of the **perineal fascia and spaces** (pouches) of the male perineum are similar to those described for the female perineum (see Case 27). For example, the membranous layer of the superficial fascia

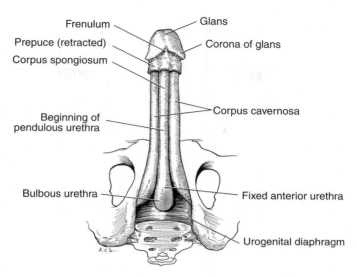

Figure 28-1. Ventral view of the penis. (*Reproduced, with permission, from Lindner HH. Clinical Anatomy. East Norwalk, CT: Appleton & Lange, 1989:498.*)

attaches to the posterior margin of the perineal membrane, the same three superficial perineal muscles are surrounded by the deep perineal fascia, and superficial and deep spaces are present. However, in the male perineum, the fatty layer of the superficial fascia is virtually absent on the penis and is replaced by **smooth (dartos) muscles** in the scrotum. The membranous layer of the superficial fascia is continuous in the penis and scrotum as the dartos fascia (Figures 28-1 and 28-2).

The **penis** is **developmentally homologous** to the **clitoris** in the female and has many anatomical similarities. However, the **urethra traverses the corpus spongiosum.**

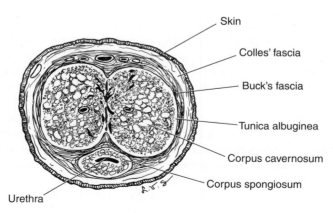

Figure 28-2. Transverse section of the penis. (*Reproduced, with permission, from Lindner HH. Clinical Anatomy. East Norwalk, CT: Appleton & Lange, 1989:500.*)

The **penis consists of root, body, and glans**, which are formed from three cylindrical bodies of erectile tissue, each surrounded by a thick fibrous capsule called the **tunica albuginea.** Paired **corpora cavernosa** attach to the posterior portion of the ischiopubic rami (the crura of the penis) and converge anteriorly at the pubic symphysis. The paired bodies fuse with each other and are flexed inferiorly. The single **corpus spongiosum** begins as an expanded region called the **bulb of the penis,** which is attached to the inferior surface of the perineal membrane, and into which the urethra passes. The crura and bulb form the root of the penis. The corpus spongiosum with the urethra within it courses anteriorly to meet and fuse with the paired corpora cavernosa and form the body of the penis. The distal portion of the corpus spongiosum is expanded as the glans, which caps the distal ends of the paired corpora cavernosa. The external urethral orifice is at the tip of the glans. The **three fused erectile bodies are surrounded by a deep (Buck) fascia,** thin loose connective tissue, and thin, somewhat pigmented skin. The glans is covered by a redundant fold of skin called the **prepuce (foreskin)** and is removed if a child is **circumcised.** The **posterior crural portion of the corpora cavernosa** are **covered with ischiocavernosus muscles,** and the **corpus spongiosum is covered by the paired bulbospongiosus muscles. Superficial transverse perineal muscles** are also present at the posterior margin of the perineal membrane and attach to the perineal body.

The **scrotum** is a sac of pigmented skin and the **dartos fascial layer,** which contains smooth muscle fibers that produce the characteristic wrinkling of the skin. The scrotum is posteroinferior to the penis and is divided into two compartments by an internal septum. Each compartment contains a **testis, epididymis, and the spermatic cord.** Each testis is ovoid with a thick fibrous capsule, the **tunica albuginea,** from which incomplete connective tissue septa divide the interior into lobules. The **lobules contain testosterone-producing interstitial cells (of Leydig)** and **coiled seminiferous tubules where spermatozoa (sperm) are produced.** The seminiferous tubules converge toward the posteriorly located mediastinum to form tubules (straight tubules, rete testes, and efferent tubules), which convey sperm to the **epididymis.** The **epididymis is the comma-shaped structure attached to the posterior surface of the testis** and is composed of the highly convoluted ductus epididymis. The testis and epididymis are surrounded by a closed, double-layered peritoneal sac embryologically derived from the **process vaginalis.** The inner portion or the visceral layer of the tunica vaginalis is applied to the surface of the testis and epididymis and is continuous posteriorly with an outer, parietal layer of the tunica vaginalis. A small cavity with lubricating fluid separates the two layers (Figure 28-3).

The epididymis is continuous inferiorly with the **ductus (vas) deferens,** which courses superiorly to enter the superficial inguinal ring. The ductus deferens along with the **testicular, deferential, and cremasteric arteries, pampiniform plexus of veins, genital branch of the genitofemoral nerve, autonomic nerve fibers, and lymphatic vessels** are components of the **spermatic cord.** The testis, epididymis, and spermatic cord are encased in three fascial layers derived from layers of the anterior abdominal wall (Table 28-1).

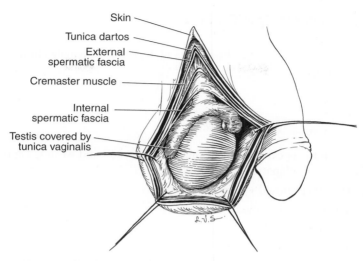

Figure 28-3. Layers of the scrotum. (*Reproduced, with permission, from Lindner HH. Clinical Anatomy. East Norwalk, CT: Appleton & Lange, 1989:501.*)

The testes are supplied by the **testicular arteries that arise from the abdominal aorta, just inferior to the renal arteries.** They course retroperitoneally to reach the deep inguinal ring, crossing anteriorly to the ureters and external iliac vessels. They traverse the inguinal canal to enter the scrotum through the **superficial inguinal ring.** Venous drainage of the testes is by the **pampiniform plexus of veins,** which follow a reverse course through the inguinal rings and canal, to become paired testicular veins near their entrance into the abdomen. Eventually, a single testicular vein is formed that drains into the **IVC on the right side** but enters the **left renal vein on the left side.** Lymphatic vessels ascend along the paths of the testicular vessels to drain lymph into **lumbar and preaortic lymph nodes** at the level of origin of the arteries. This high abdominal position of arterial origin and lymphatic drainage reflects the embryological site where the testes were formed.

Table 28-1 • ORIGINS OF SPERMATIC CORD COVERINGS	
Fascia	Abdominal Layer of Origin
Internal spermatic fascia	Transversalis fascia
Cremasteric fascia and muscle	Internal abdominal oblique muscle
External spermatic fascia	External abdominal oblique muscle

COMPREHENSION QUESTIONS

28.1 Which of the following is the male homologue of the female clitoris?
   A. Epididymis
   B. Vas deferens
   C. Penis
   D. Scrotum

28.2 The scrotum appears to have a slightly pigmented and wrinkled appearance. What is the explanation for this appearance?
   A. Hyperkeratinized squamous epithelium
   B. The tunica albuginea
   C. The dartos fascia
   D. The pampiniform plexus

28.3 An 18-year-old man is noted to have probable testicular cancer. He undergoes surgery. After incising the scrotum, the surgeon contemplates the approach to the parenchyma of the testes. Through which layer must the surgeon incise to reach the testicular parenchyma?
   A. Buck fascia
   B. Tunica albuginea
   C. Dartos fascia
   D. Scarpa fascia

28.4 A 7-year-old male comes in for a routine physical examination. The pediatrician notices that the right testis is enlarged and without tenderness. Transillumination reveals clear fluid which is present around the right testis. This fluid most likely occupies which space?
   A. Tunica albuginea
   B. External spermatic fascia
   C. Tunica vaginalis
   D. Cremasteric layer

ANSWERS

28.1 **C.** The penis in the male is the homologue to the clitoris in the female.

28.2 **C.** The dartos fascia, which consists of smooth muscle, gives the scrotum its characteristic slightly pigmented and wrinkled appearance.

28.3 **B.** Each testis is surrounded by a thick capsule, the tunica albuginea.

28.4 **C.** This patient most likely has a **hydrocele**, which is a fluid collection in the tunica vaginalis. This is a congenital condition formed when the testis descends through the inguinal canal together with some peritoneum. Peritoneal fluid sometimes accumulates in this space.

ANATOMY PEARLS

▶ The root of the penis is defined as the crura and the bulb.

▶ The cremasteric muscle, which causes elevation of the testes in the cremasteric reflex, is innervated by the genital branch of the genitofemoral nerve.

▶ The testicular artery arises from the aorta just inferior to the renal arteries.

▶ The right testicular vein drains into the IVC, whereas the left one drains into the left renal vein.

REFERENCES

Gilroy AM, MacPherson BR, Ross LM. *Atlas of Anatomy,* 2nd ed. New York: Thieme Medical Publishers; 2012:251, 257, 259–260.

Moore KL, Dalley AF, Agur AMR. *Clinically Oriented Anatomy,* 7th ed. Baltimore, MD: Lippincott Williams & Wilkins; 2014:206–210, 215.

Netter FH. *Atlas of Human Anatomy,* 6th ed. Philadelphia, PA: Saunders; 2014: plates 358–360, 365.

A 50-year-old female who has borne five children complains that she has noticed vaginal spotting of blood after intercourse for approximately the past 6 months. More recently, she has had a foul-smelling vaginal discharge and indicates that her left leg seems larger than her right one. She previously had syphilis. She has smoked one pack of cigarettes per day for 20 years. Examination of her back shows left flank tenderness. The circumferences of her left thigh and calf are larger than those of the right. Pelvic examination shows normal female external genitalia and a 3-cm growth on the surface on the left lip of the uterine cervix.

▶ What is the most likely diagnosis?
▶ What is the applied clinical anatomy for this condition?

ANSWER TO CASE 29:

Metastatic Cervical Cancer With Ureter Obstruction

Summary: A 50-year-old female who has borne five children complains of a 6-month history of spotting after intercourse and foul-smelling vaginal discharge. She has had syphilis and is a smoker. Left costovertebral tenderness is present, and the left lower limb is swollen. Speculum examination of the uterine cervix shows a 3-cm growth of the left lip of the cervix.

- **Most likely diagnosis:** Metastatic cervical cancer

- **Applied anatomy for this condition:** Extension of the tumor to obstruct the left ureter and metastasis to iliac lymph nodes

CLINICAL CORRELATION

This patient's age, multiple pregnancies, and histories of smoking and a sexually transmitted disease are risk factors for cervical cancer. Vaginal spotting after intercourse is a common presenting sign for cervical cancer in a sexually active woman. Cervical cancer typically arises at the squamocolumnar epithelial junction, and the foul-smelling discharge suggests necrosis of a portion of this large tumor. Such a tumor can spread inferiorly to involve the vagina or laterally into the region of the transverse cervical (cardinal) ligament and can obstruct the ureter, which passes through the ligament. Further growth may reach the lateral pelvic wall. Involvement of iliac lymph nodes, in particular the external iliac nodes, may inhibit lymphatic drainage of the lower limb with resultant edema. Bilateral ureteral obstruction can lead to uremia, the most common cause of death in this disease. Radiotherapy is the primary treatment for advanced cervical cancer.

APPROACH TO:

The Internal Female Genital System I

OBJECTIVES

1. Be able to describe the anatomy of the ovaries, uterine tubes, uterus, and upper vagina, including changes in their epithelial lining

2. Be able to describe the anatomy of the lateral uterine support structures and related organs

3. Be able to draw the lymphatic drainage of the uterus and upper vagina

DEFINITIONS

POSTCOITAL SPOTTING: Vaginal bleeding after sexual intercourse, usually due to friable cervical tissue, that may be a sign of cervical inflammation or cancer

CERVICAL DYSPLASIA: Premalignant condition of the cervical epithelium usually induced by human papilloma virus, which over time may evolve into cervical cancer

CERVICAL CYTOLOGY: Method of studying cells obtained by scrapings from the cervix

COLPOSCOPIC EXAMINATION: Method of visually examining the cervix with a binocular magnifying device, usually with the addition of acetic acid to locate areas of cervical dysplasia

DISCUSSION

The **uterus** or womb is a thick-walled, hollow, **pear-shaped**, pelvic organ. Its main parts are the **body and cervix.** The **fundus** is the superior portion of the body between the openings of the uterine tubes, and the **isthmus** is the narrowed inferior portion of the body at its junction with the cervix. The narrow **uterine cervix** protrudes into the anterior wall of the upper vagina. The lumen of the cervix is the **cervical canal**; its superior part opens into the uterine cavity as the internal os, and its inferior part opens into the vagina as the external os. The uterus is usually angled anteriorly in relation to the vagina, or **anteverted**, and the body and cervix of the uterus are flexed anteriorly with respect to each other, or **anteflexed.** This places the body of the uterus superior to the urinary bladder, often deforming it on cystograms. Posterior to the cervix is the rectum. The **vagina**, a tubular structure that is closed anteroposteriorly, begins at the **vestibule** and is directed posterosuperiorly to the level of the cervix. The protrusion of the cervix into the anterior wall of the vagina creates a circumferential gutter around the cervix, which, although a continuous space, is typically referred to as the **anterior, posterior, or lateral fornix.** The **urethra** is embedded in the **anterior wall of the vagina.** The columnar epithelium, which lines the uterine cavity and cervical canal, changes to a nonkeratinized stratified squamous epithelium at the margins of the external os. This type of epithelium covers the external surface of the cervix and lines the vagina (Figure 29-1).

The **uterine (fallopian) tubes** extend posterolaterally from the superolateral region of the uterus, the uterine horns. The uterine tubes are divided, from medial to lateral, into four regions: a **uterine or intramural portion** within the wall of the uterus, the narrowest portion or **isthmus**, the widest portion or **ampulla**, and the funnel-shaped **infundibulum.** The lumen of the infundibulum opens into the abdominal cavity, and its margin is arranged in a series of finger-like structures called **fimbriae**, one of which is usually attached to the ovary. The **female gonads, the ovaries**, lie close to the lateral pelvic wall, just inferior to the pelvic brim. Each almond-shaped ovary is supported by a **suspensory (infundibulopelvic) ligament**, which consists of the peritoneally covered ovarian vessels, an ovarian ligament, a derivative of the proximal portion of the embryonic gubernaculum, and the mesovarium portion of the broad ligament.

The uterus, uterine tubes, and ovary are draped by a mesentery, the **broad ligament**, which passes from the sides of the uterus to the lateral pelvic wall to divide the pelvic cavity into anterior and posterior compartments. The broad ligament has three subdivisions: a shelflike portion derived from the posterior layer of the broad

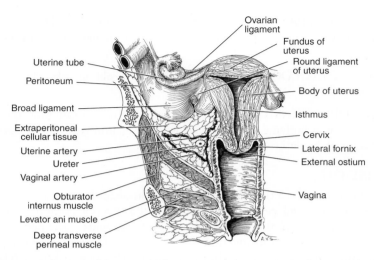

Figure 29-1. Frontal section of the uterus and vagina. (*Reproduced, with permission, from Lindner HH. Clinical Anatomy. East Norwalk, CT: Appleton & Lange, 1989:514.*)

ligament that attaches to the ovary, called the **mesovarium;** the portion of the broad ligament superior to the mesovarium that attaches to the uterine tube, called the **mesosalpinx;** and the portion from the uterus to the lateral pelvic wall, called the **mesometrium.** The continuation of peritoneum from the anterior surface of the uterus onto the anterior placed urinary bladder creates the **uterovesicular pouch.** Similarly, the continuation of peritoneum from the posterior uterine surface onto the anterior surface of the rectum creates the **rectouterine pouch** (of Douglas), the **most inferior recess of the abdominopelvic cavity in the female.**

The uterus and uterine tubes are covered with a layer of visceral peritoneum, but the ovary is not; it is covered instead by a cuboidal germinal epithelium. The ovarian ligament is a cordlike structure between the layers of the mesovarium that extends from the uterine pole of the ovary to the uterine horn. Its continuation anteriorly to and through the deep inguinal ring and inguinal canal to the labia majora is the **round ligament of the uterus** (also derived from the gubernaculum). Beneath the peritoneum of the pelvic floor, paired condensations of connective tissue, the **uterosacral ligaments,** pass from the uterine cervix to the sacrum. An additional pair of condensation passes from the cervix to the lateral pelvic wall, the **transverse cervical (cardinal) ligaments.** The transverse cervical ligaments lie in the base of the mesometrium, and the uterine vessels lie within or very close to these ligaments. The ureters coursing anteromedially on their way to the urinary bladder **pass inferiorly to the uterine vessels (mnemonic:** "water under the bridge") and continue anteriorly, approximately 2 cm laterally to the uterine cervix.

The **blood supply to the uterus** consists primarily of the **paired uterine arteries** and the **ovarian arteries.** The uterine arteries arise from the internal iliac arteries and traverse through the transverse cervical (cardinal) ligaments. The fundus (top) of the uterus is supplied mainly by the ovarian arteries, which arise from the abdominal aorta. **Lymphatic drainage** from the fundus and body of the uterus is to the **lumbar abdominal nodes and the external iliac nodes.** The **cervical lymph**

drainage is primarily to **external iliac nodes,** but some lymph drains to **internal iliac and sacral nodes.** Drainage from the upper vagina is similar to that of the cervix, to the external and internal iliac lymph nodes.

COMPREHENSION QUESTIONS

29.1 A 31-year-old woman is in her physician's office for a fitting for an intrauterine contraceptive device. The physician performs a pelvic examination to ensure that the device is placed in the correct direction. The physical examination shows that the uterine body is tipped toward the rectum and that the uterine fundus is tipped anteriorly. Which of the following describes the position of the uterus?

    A. Anteverted, anteflexed

    B. Anteverted, retroflexed

    C. Retroverted, anteflexed

    D. Retroverted, retroflexed

29.2 A 45-year-old woman is having significant uterine bleeding from uterine fibroids. The radiologist performs an embolization procedure of the uterine arteries. Through which of the following structures do the uterine arteries traverse?

    A. Transverse cervical (cardinal) ligaments

    B. Uterosacral ligaments

    C. Vesicouterine fold

    D. Anterior vaginal fornix

29.3 A 20-gauge spinal needle is placed through the vagina to assess whether there is blood in the peritoneal cavity. Which of the following describes the most dependent part of the peritoneum or pelvis?

    A. Vesicouterine fold

    B. Pararectal space

    C. Paravesical space

    D. Rectouterine pouch (of Douglas)

29.4 A 42-year-old woman is undergoing a total abdominal hysterectomy due to leiomyomata of the uterus that has caused significant abnormal vaginal bleeding. During the surgery, the surgeon locates the left ureter to ensure its safety prior to clamping the uterine artery. The ureter is found at the pelvic brim. In this area, the left ureter is located immediately lateral to the

    A. Left ovarian vein

    B. Left external iliac artery

    C. Abdominal aorta

    D. Left internal iliac artery

    E. Left uterine artery

    F. Left renal artery

ANSWERS

29.1 **C.** "Version" refers to the relation between the cervix and uterine body, whereas "flexion" denotes the relation between the uterine body and the uterine fundus (top). Thus, this uterus is retroverted and anteflexed.

29.2 **A.** The uterine arteries travel through the transverse cervical ligaments.

29.3 **D.** The most dependent region of the pelvis is the rectouterine pouch of Douglas. The procedure described is called a **culdocentesis**, in which the spinal needle is placed through the posterior vaginal fornix.

29.4 **D.** One of the key surgical anatomical landmarks for the ureter is at the pelvic brim, in which the ureter crosses medially at the bifurcation of the common iliac artery. At this location, the ureter is medial to the ovarian vessels and lateral to the internal iliac artery and vein. From this location, the ureters travel more medially, under the uterine artery, to the bladder.

ANATOMY PEARLS

▶ The posterior vaginal fornix is in close relation to the rectouterine pouch (of Douglas), the most inferior portion of the abdominopelvic cavity in the female.

▶ The suspensory ligament of the ovary contains the ovarian vessels.

▶ After passing inferiorly to the uterine vessels, the ureters course medially and lie approximately 2 cm laterally to the uterine cervix.

▶ Lymph from the uterine cervix and upper vagina drains primarily to the external iliac node group.

REFERENCES

Gilroy AM, MacPherson BR, Ross LM. *Atlas of Anatomy*, 2nd ed. New York, NY: Thieme Medical Publishers; 2012:237, 240, 243, 245.

Moore KL, Dalley AF, Agur AMR. *Clinically Oriented Anatomy*, 7th ed. Baltimore, MD: Lippincott Williams & Wilkins, 2014:385–389, 395.

Netter FH. *Atlas of Human Anatomy*, 6th ed. Philadelphia, PA: Saunders; 2014: plates 340–342, 346, 352.

A pregnant 19-year-old female who has borne one healthy child is being seen at 7 weeks' gestation based on her last menstrual period and her complaints of vaginal spotting and lower abdominal pain. She denies the passage of any tissue through the vagina, trauma, or recent intercourse. Her medical history is significant for a pelvic infection approximately 3 years previously. On examination, her blood pressure is 90/60 mmHg, heart rate is 110 beats/min, and temperature is within normal limits. The abdomen is normal, and bowel sounds are present and normal. On pelvic examination, the external genitalia and uterus palpate as normal. There is moderate right adnexal tenderness with palpation. Quantitative human β-corticotropin gonadotropin is 2300 mIU/mL, and a transvaginal sonogram displays an empty uterus and some free fluid in the cul-de-sac.

▶ What is the most likely diagnosis?
▶ What is the cause of the hypotension?

ANSWER TO CASE 30:
Ectopic Pregnancy

Summary: A 19-year-old female who has borne one child is seen at 7 weeks' gestation by last menstrual period and vaginal spotting. She has a history of a pelvic infection. Her blood pressure is 90/60 mmHg, heart rate is 110 beats/min, and the abdomen is mildly tender. Pelvic examination shows a normal uterus and some moderate adnexal tenderness. Quantitative human β-corticotropin gonadotropin is 2300 mIU/mL, and transvaginal sonogram shows an empty uterus and some free fluid in the cul-de-sac.

- **Most likely diagnosis:** Ectopic pregnancy

- **Cause of the hypotension:** Ruptured ectopic pregnancy in the uterine tube with bleeding into the abdominal cavity

CLINICAL CORRELATION

An ectopic pregnancy results when a blastocyst implants outside the lumen of the uterus. The vast majority of ectopic pregnancies occur in the uterine tube (95 to 97 percent), in either the ampulla, the usual site of fertilization; or the isthmus, the narrowest portion. Any condition that might prevent or delay transport of the zygote to the uterus may cause an ectopic tubal pregnancy, and this patient's history of a pelvic infection (pelvic inflammatory disease) is a risk factor. Tubal ectopic pregnancies will usually rupture during the first 8 weeks of pregnancy, typically resulting in abortion of the embryo and intraabdominal hemorrhage, with resultant hypotension and tachycardia. Tubal pregnancy in the narrow isthmus tends to rupture sooner than those in the ampulla and produce greater hemorrhage than implantation in the ampulla. Blastocysts implanted in the ampulla may be expelled into the abdominal cavity, where they may reimplant on the surface of the ovary, the peritoneum of the rectouterine pouch (of Douglas), mesentery, or organ surface. Severe hemorrhage typically results from an abdominal ectopic pregnancy, and the resulting hypotension may be emergent. The free fluid seen on ultrasound is blood that has resulted from the ruptured ectopic pregnancy.

APPROACH TO:
The Internal Female Genital System II

OBJECTIVES

1. Be able to describe the anatomy of the uterine tubes

2. Be able to draw the blood supply to the ovaries, uterine tubes, and uterus

DEFINITIONS

ECTOPIC PREGNANCY: Pregnancy outside of the normal endometrial implantation site, usually involving the fallopian tubes

HEMOPERITONEUM: Blood collecting inside the peritoneal cavity, usually leading to abdominal pain and irritation to the intestines

HUMAN CHORIONIC GONADOTROPIN: Glycoprotein molecule produced by the trophoblastic cells of the pregnancy

DISCUSSION

The **uterine (fallopian) tubes** (see Case 29) extend posterolaterally from the uterine horns and are divided, from medial to lateral, into **four regions.** The **uterine or intramural portion lies within the wall of the uterus.** The **narrowest portion, or isthmus**, lies just laterally to the uterine horns. More laterally, **the widest and longest portion of the tube is the ampulla.** This is the **usual site of fertilization.** The most **lateral portion or infundibulum is funnel-shaped.** The lumen of the infundibulum faces posteriorly into the abdominal cavity, inferior to which is the rectouterine pouch (of Douglas). The margin of the infundibulum is arranged in a series of fingerlike structures called **fimbriae**, one of which is usually attached to the ovary. This attachment helps keep the infundibulum in close anatomical relation to the ovary, which, in turn, helps ensure that an ovulated egg will enter the lumen of the tube. The uterine tube is supported by the **mesosalpinx** portion of the broad ligament (Figure 30-1).

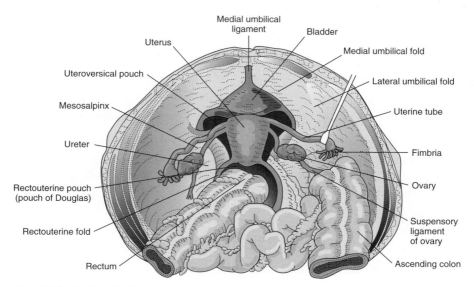

Figure 30-1. The female pelvis and internal organs (superior view). (*Reproduced, with permission, from DeCherney AH, Nathan L. Current Obstetric and Gynecology Diagnosis and Treatment, 9th ed. New York: McGraw-Hill, 2003:33.*)

The ovaries, uterine tubes, and fundus of the uterus are supplied by the **ovarian arteries**, which arise from the abdominal aorta just inferior to the renal arteries (in a manner similar to that described for the testicular arteries). The arteries descend, **crossing the ureters anteriorly**, and also cross the iliac vessels anteriorly at the pelvic brim. The ureters lie just medial at the pelvic brim. The arteries enter the lateral pole of each ovary, supply it, and continue medially between the layers of the mesosalpinx, close to its attachment to the uterine tube. Each artery supplies the tube, continues on to supply the fundus of the uterus, and anastomoses with the artery from the opposite side. The **isthmus and uterine portions of the tube also receive blood from ascending branches of the uterine arteries, which anastomose with the ovarian artery.** This accounts for the increased hemorrhage with a ruptured tubal pregnancy of the isthmus. Venous drainage from these structures is primarily through the ovarian veins, which empty into the IVC on the right side and into the left renal vein on the left side.

COMPREHENSION QUESTIONS

30.1 A 22-year-old woman is noted during surgery to have a 3-cm ectopic pregnancy involving the ampulla of the fallopian tube. Which of the following best describes this location of the tube?

A. Portion within the muscle of the uterus

B. Portion that is narrowest and mobile

C. Portion that begins to widen distally and is the longest portion of the tube

D. Portion with fingerlike projections

30.2 Bilateral oophorectomy is performed in a woman who had ovarian cancer. To accomplish this procedure, the ovarian arteries were ligated. Which of the following describes the anatomy of the ovarian vessels?

A. Right ovarian artery arises from the right renal artery.

B. Right ovarian vein drains into the vena cava.

C. Left ovarian artery arises from the left internal iliac artery.

D. Left ovarian vein drains into the vena cava.

30.3 A 3-cm ectopic pregnancy of the isthmus of the left tube is noted to have ruptured, leading to hemorrhage. The blood noted arises principally from which of the following?

A. Uterine artery

B. Ovarian artery

C. Uterine and ovarian arteries

D. Neither the uterine nor the ovarian arteries

ANSWERS

30.1 **C.** The ampulla of the tube, which is the most common location of ectopic pregnancies, is the part of the tube that begins to widen at the distal end of the tube.

30.2 **B.** Both ovarian arteries arise from the abdominal aorta. The right ovarian vein drains to the vena cava, whereas the left ovarian vein drains into the left renal vein.

30.3 **C.** The uterine artery (ascending branch) and the ovarian artery anastomose to provide blood supply within the mesosalpinx to the tube.

ANATOMY PEARLS

▶ The usual site of fertilization is the ampulla of the uterine tube.

▶ The posteriorly facing ostium of the tube accounts for abdominal ectopic pregnancies usually occurring in the rectouterine pouch.

▶ The ovarian artery supplies the ovary, uterine tube, and fundus of the uterus. Anastomosis with the uterine artery occurs in the region of the isthmus.

REFERENCES

Gilroy AM, MacPherson BR, Ross LM. *Atlas of Anatomy*, 2nd ed. New York, NY: Thieme Medical Publishers; 2012:230, 237, 243, 247.

Moore KL, Dalley AF, Agur AMR. *Clinically Oriented Anatomy*, 7th ed. Baltimore, MD: Lippincott Williams & Wilkins; 2014:382–385, 392.

Netter FH. *Atlas of Human Anatomy*, 6th ed. Philadelphia, PA: Saunders; 2014: plates 340–342, 350–353.

A 63-year-old male complains of a 6-month history of difficulty voiding and feeling that he cannot empty his bladder completely. After voiding, he often feels the urge to urinate again. He denies urethral discharge or burning with urination. He has had mild hypertension and takes a thiazide diuretic. His only other medication has been ampicillin for two urinary tract infections during the previous year. On examination, his blood pressure is 130/84 mm Hg, his pulse rate is 80 beats/min, and he is without fever (afebrile). The heart and lung examinations are normal, and the abdominal examination shows no masses.

▶ What is the most likely diagnosis?
▶ What is the anatomical explanation for the patient's symptoms?

ANSWER TO CASE 31:

Benign Prostatic Hyperplasia

Summary: A 63-year-old male who has hypertension complains of a 6-month diffi-culty in voiding and the sensation that he cannot empty his bladder completely. He has had two episodes of urinary tract infections but denies dysuria (uncomfortable burning sensation on urination) or urethral discharge.

- **Most likely diagnosis:** Benign prostatic hyperplasia

- **Anatomical basis for the symptomatology:** Compression of the bladder neck or the prostatic urethra

CLINICAL CORRELATION

The **prostate gland** is the **largest of the male accessory sex glands**, and its secre-tions contribute to semen. This **encapsulated gland** is located in the pelvis, between the neck of the bladder and the sphincter urethrae muscle, and **surrounds the first part of the male urethra,** called the **prostatic urethra.** Enlargement of the prostate, **benign prostatic hyperplasia** (BPH), is a common condition in men 50 years and older and appears to depend on age and hormone level. A **prostate-specific antigen** blood test and a digital rectal examination (DRE) would be done to evaluate the gland's size and the presence of nodularity, which might suggest carcinoma. Initial treatment after a confirmed diagnosis of BPH is often medical, with a medication such as a **5-α-reductase inhibitor,** which relaxes the smooth muscle within the stroma of the gland and thus increases urethral diameter. Other medications block the effects of testosterone metabolites on gland tissue, resulting in involution of gland tissue. In advanced cases, a surgical transurethral resection of the prostate may be required. Although no direct relation between BPH and prostate malignancy has been proved, both conditions occur in the same age group.

APPROACH TO:

Male Internal Genitalia

OBJECTIVES

1. Be able to describe the anatomy of the internal male genital organs: ductus deferens, seminal gland, ejaculatory duct, prostate gland, and bulbourethral glands

2. Be able to describe the anatomy of the male urethral tract

DEFINITIONS

PROSTATIC HYPERPLASIA: Benign enlargement of the prostate gland that, because of the capsule surrounding it, impinges on the urethra

URINARY HESITANCY: Abnormally long period required to initiate a stream of urine

TRANSURETHRAL RESECTION OF THE PROSTATE: Procedure in which the surgeon excises prostatic tissue from the prostatic urethra in an effort to relieve obstruction

DISCUSSION

The **paired ductus deferenses traverse the inguinal canal** and enter the abdomen through the **deep inguinal rings**, where they retain a retroperitoneal position. They cross the external iliac vessels and superolateral surface of the bladder, continue superior to the ureters entering the bladder (mnemonic: "water under the bridge"), and reach the posterior surface of the bladder, just anterior to the rectal vesicular pouch. The terminal portion of the ductus is dilated to form an **ampulla of ductus deferens.** Lateral to the two ampullae are the diagonally positioned, paired **seminal(s) gland(s).** These accessory sex glands produce an **alkaline component** of semen, which neutralizes the usual acid environment in the vagina. The duct of each **seminal gland unites with the ductus deferens on each side to form the paired ejaculatory ducts,** which course anteroinferiorly through the prostate gland to open on the elevated seminal colliculus on the posterior wall of the prostatic urethra (Figure 31-1).

The **prostate gland is the largest of the accessory sex glands,** an inverted pyramid about the **size of a walnut.** The **base is located inferior to the neck of the bladder,** and the apex rests on the sphincter urethral muscle. The prostate has a **thick fibrous capsule** surrounded by a fibrous sheath that is continuous with the puboprostatic ligaments. The levator ani muscle supports the gland inferolaterally, and the anterior surface is covered by fibers of the sphincter urethral muscle. The prostate is anatomically divisible into **four lobes.** The **anterior lobe** lies anterior to the urethra and is a superior fibromuscular continuation of the sphincter urethral muscle. The **posterior lobe** is midline, posterior to the urethra, and palpable by digital rectal exam (DRE). The **lateral lobes** on each side of the **posterior lobe** form the largest part of the gland and are also palpable by DRE. The **middle lobe is the wedge-shaped superior portion of gland between the urethra** and the obliquely oriented ejaculatory ducts and is **closely related to the neck of the bladder. Enlargement** of the **middle lobe** (as in BPH) causes pressure on the neck of the bladder. The multiple ducts of the prostate open onto the posterior wall of the prostatic urethra and constitute a major component of semen. The paired **bulbourethral glands** are pea-size glands embedded in the sphincter urethral muscle, posterolateral to the membranous urethra. The ducts of each gland empty into the proximal part of the spongy (penile) urethra in the bulb of the penis. **Their mucous secretions lubricate the urethra during erection.**

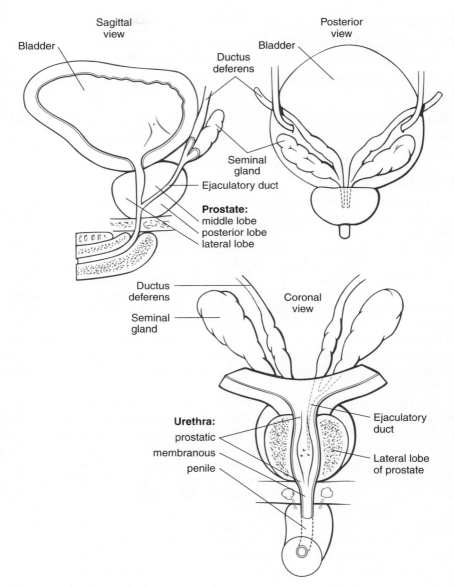

Figure 31-1. The prostate gland.

The **male urethra** is divided **into four parts.** The **preprostatic urethra** is a short continuation of the bladder neck. The **prostatic urethra**, the widest part, passes through the prostate gland, somewhat closer to its anterior surface. The posterior wall is elevated as a **fusiform ridge** called the **seminal colliculus**, on which are found the openings of the prostatic utricle (an embryonic remnant) and the paired ejaculatory ducts. The grooved portions of the urethra on each side of the colliculus are the prostatic sinuses, which contain the openings of the prostatic gland ducts.

The third part, the **membranous urethra,** is surrounded by the sphincter urethral muscle, or the external urethral (voluntary) sphincter. The fourth and longest part is the **spongy (penile) urethra**, which traverses the **corpus spongiosum** and terminates at the external urethral orifice on the tip of the glans penis. As the urethra enters the bulb of the penis, it widens to form the bulbar fossa into which open the ducts of the **bulbourethral glands**. The urethra widens again just proximal to the external orifice as the **navicular fossa**.

COMPREHENSION QUESTIONS

31.1 A 66-year-old man complains of difficulty voiding and is noted to have probable BPH. Which of the following prostatic lobes is likely to be responsible for these symptoms?

A. Anterior lobe

B. Posterior lobe

C. Lateral lobe

D. Middle lobe

31.2 A 48-year-old man is undergoing cystoscopic examination. As the cystoscope is placed into the urethra through the penile portion, which of the following tissues surrounds the urethra?

A. Prostate

B. Corpus spongiosum

C. Seminal colliculus

D. Sphincter urethral muscles

31.3 A police detective takes a scraping of some stains to be examined for alkaline phosphatase to assess whether these might be ejaculate. What is the source of alkaline phosphatase in the semen?

A. Prostatic gland

B. Bulbourethral glands

C. Seminal gland

D. Seminal colliculus apparatus

ANSWERS

31.1 **D.** The middle lobe of the prostate is the part through which the urethra traverses and may be obstructed by BPH.

31.2 **B.** The longest portion of the urethra is the penile urethra, which traverses through the corpus spongiosum.

31.3 **C.** The seminal gland are the source of the alkaline phosphatase in the semen. The alkalinity helps to neutralize the acidity of the vagina.

ANATOMY PEARLS

▶ The posterior and lateral lobes of the prostate are palpable by DRE.

▶ The middle lobe may press on the bladder neck in BPH.

▶ The sphincter urethral muscle extends superiorly to cover the anterior surface of the prostate.

▶ The bulbourethral glands lie adjacent to the membranous urethra, but their ducts open into the proximal spongy urethra.

REFERENCES

Gilroy AM, MacPherson BR, Ross LM. *Atlas of Anatomy*, 2nd ed. New York, NY: Thieme Medical Publishers; 2012:252–253.

Moore KL, Dalley AF, Agur AMR. *Clinically Oriented Anatomy*, 7th ed. Baltimore, MD: Lippincott Williams & Wilkins; 2014:376–379, 381.

Netter FH. *Atlas of Human Anatomy*, 6th ed. Philadelphia, PA: Saunders; 2014: plates 361–363.

A 45-year-old female underwent surgical removal of the uterus (total hysterec-tomy) for symptomatic endometriosis 2 days previously. She complains of right back and flank tenderness. On examination, her temperature is 102°F, heart rate is 100 beats/min, and blood pressure is 130/90 mmHg. The heart and lung examina-tions are normal. Her abdomen is slightly tender diffusely, but bowel sounds are normal. The surgical incision appears within normal limits. There is exquisite right costovertebral angle tenderness on palpation. Ultrasound of the kidneys shows marked dilation of the right renal collecting system and dilation of the right ureter.

► What is the most likely diagnosis?
► What is the anatomical explanation for this condition?

ANSWER TO CASE 32:
Ureteral Injury at Surgery

Summary: A 45-year-old female who underwent total abdominal hysterectomy for symptomatic endometriosis 2 days previously has fever of 102°F and exquisite right flank tenderness at the costovertebral angle. The surgical incision appears normal.

- **Most likely diagnosis:** Injury to the right ureter

- **Anatomical explanation for this condition:** Probable ligation of the right ureter as it passes inferiorly to the uterine artery within the transverse cervical (cardinal) ligament of the uterus

CLINICAL CORRELATION

Approximately one-half the length of the ureter is located in the pelvis. It is at risk for injury at three pelvic sites during a hysterectomy. If the patient's ovaries are also removed (oophorectomy) at the time of the hysterectomy, the ureter is at risk where it crosses the common or external iliac vessels to enter the pelvis just medial to the ovarian vessels. The ureter is especially at risk deeper in the pelvis as it courses toward the urinary bladder inferior to the uterine vessels. Lateral extension of uterine pathology into the transverse cervical ligament increases the risk. The third site at which the ureter is at risk is as it passes laterally to the uterine cervix before its entrance into the urinary bladder. Hydronephrosis and/or hydroureter results from ureteral injury, and cystoscopic stent passage is often attempted first to relieve the obstruction, if possible.

APPROACH TO:
The Ureters

OBJECTIVES

1. Be able to draw the abdominal and pelvic courses of the ureter

2. Be able to describe the sites at which the ureter is anatomically narrowed and at risk during surgery

3. Be able to describe the blood supply to the ureter

DEFINITIONS

HYSTERECTOMY: Surgical removal of the uterus.

URETERAL INJURY: Ligation, laceration, or denuding the ureter leading to ischemia. Ureteral obstruction can also occur from "kinking" of the ureter.

INTRAVENOUS PYELOGRAM: Intravenous dye is injected, and a series of radiographs are taken that incorporate the kidneys, ureter, and bladder. This procedure allows delineation of the anatomical structures and the function of the kidneys.

DISCUSSION

Each **ureter** is an inferior continuation of the renal pelvis, and the ureteropelvic junction is at the inferior margin of the hilum of the kidney. One-half the total length of the ureter is abdominal, and the remaining half is located in the pelvis. The abdominal ureter descends **retroperitoneally** on the anterior surface of the **psoas muscle,** and at about its midpoint it is **crossed anteriorly by the gonadal arteries (testicular/ovarian).** The left ureter lies at the apex of the mesosigmoid. It enters the pelvis by crossing anterior to the external iliac artery (it may cross the common iliac bifurcation somewhat medially). In females, the ovarian vessels lie just lateral to the ureters as they enter the pelvis. After entering the pelvis, each ureter passes inferoposteriorly, anterior to the internal iliac vessels, to above the ischial spines. The ureters then course anteromedially to the posterior bladder wall. In this course in a female, the **ureters pass inferior to the uterine vessels,** reaching the uterus from the lateral pelvic wall (mnemonic: "water under the bridge"), and lie approximately **1 cm lateral to the uterine cervix.** Externally, the ureters enter the bladder approximately 5 cm apart but course obliquely through the bladder wall such that their internal openings are only 2.5 cm apart (Figure 32-1).

[Each ureter is anatomically narrowed in **three locations:** the **uteropelvic junction (renal pelvis),** where the ureters cross the **external iliac vessels (pelvic brim),** and as the ureters obliquely **traverse the bladder wall (ureterovessical junction). Renal stones may lodge at these narrowed points.**] Calculi can form anywhere along the trajectory of each ureter: within the kidney (as **renal stones**), within the ureter proper (**ureterolithiasis**), or within the narrowest segment of the ureter, inside the bladder (as **gallstones**). The pelvic portion of the ureters is at surgical risk, especially

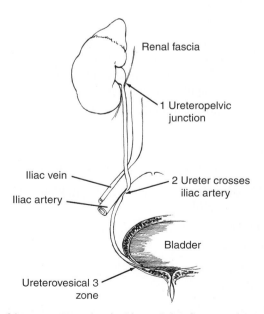

Figure 32-1. Course of the ureter. (*Reproduced, with permission, from Tanagh EA, McAninch JW, eds. Smith's General Urology, 12th ed. East Norwalk, CT: Appleton & Lange, 1988.*)

in females during **hysterectomy** procedures. If an oophorectomy is **performed with the hysterectomy, the ovarian vessels must be ligated, and each ureter lies just medial to these vessels within the suspensory ligament of the ovary.** They may be inadvertently clamped, ligated, or divided at this site. The ureters are at risk as they pass inferiorly to the uterine vessels, in or adjacent to the **transverse cervical (cardinal) ligament**, where they may also be inadvertently clamped, ligated, or divided. The ureters are at risk in a vaginal hysterectomy as they course just laterally to the uterine cervix.

The arterial blood supply of the ureters is likely to originate from any nearby artery, and its chief supply is derived from **ureteral branches from the aorta and the renal, gonadal, common and internal iliac, vesical and uterine arteries.** Ureteric branches reach the ureters from their medial side and divide into ascending and descending branches.

COMPREHENSION QUESTIONS

32.1 A 39-year-old woman complains of hematuria and significant flank tenderness. She has a history of kidney stones. A CT scan depicts the abdominal portion of the ureter lying anterior to a muscle. Which of the following is most likely to be the name of this muscle?

A. Psoas

B. Serratus anterior muscle

C. Obturator muscle

D. Rectus muscle

E. External oblique muscle

32.2 A dissection of the ureter is accomplished to excavate a large retroperitoneal mass. In isolating the ureter, the surgeon is attempting to ensure that the blood supply to the ureter is not disrupted. Which of the following best describes the arterial supply to the ureter?

A. Ureteral artery arising from the abdominal aorta

B. Ureteral artery arising from the external iliac artery

C. Ureteral artery arising from the internal iliac artery

D. No specific artery, but rather small branches from the nearby arteries

32.3 A 30-year-old woman is noted to have an absent kidney. Which of the following findings is she also likely to have?

A. Absent unilateral ovary

B. Unicornuate uterus

C. Imperforate hymen

D. Inguinal hernia

ANSWERS

32.1 **A.** The abdominal ureter lies anterior to the psoas muscle.

32.2 **D.** The ureter does not have any specific artery supplying it but rather has small branches from the nearby arteries such as the aorta and renal, gonadal, common and internal iliac, vesical and uterine arteries.

32.3 **B.** The urinary and paramesonephric ducts are in close proximity anatomically and functionally during embryologic development. Thus, a congenital abnormality in the kidney or ureter often is associated with an abnormality of the ipsilateral tube, uterine horn, or cervix. A unicornuate uterus is a condition in which one mullerian duct does not form or descend normally in embryonic development, leaving only one uterine "horn." The distal vagina, vulva, and ovary are of different embryonic origin.

ANATOMY PEARLS

▶ The ureters are retroperitoneal along their entire length.

▶ The ureters are narrowed at three sites: ureteropelvic junction, the crossing of the external iliac artery, and the location where they pass through the bladder wall.

▶ The ureters are at risk at three pelvic sites: where the ovarian vessels lie just lateral, where they pass inferior to the uterine vessels, and just lateral to the uterine cervix.

▶ Ureteral arterial branches reach the ureter from their medial side.

REFERENCES

Gilroy AM, MacPherson BR, Ross LM. *Atlas of Anatomy*, 2nd ed. New York, NY: Thieme Medical Publishers; 2012:236–237.

Moore KL, Dalley AF, Agur AMR. *Clinically Oriented Anatomy*, 7th ed. Baltimore, MD: Lippincott Williams & Wilkins, 2014:292–294, 363–364, 373.

Netter FH. *Atlas of Human Anatomy*, 6th ed. Philadelphia, PA: Saunders; 2014: plates 313–314, 316–318.

A 54-year-old man complains of lower back pain that radiates down the back of his right leg. He states that the pain increases when he coughs or lifts objects but decreases when he lies down. He denies having experienced trauma to his back. On examination, the strength and sensation of his lower extremities are normal. During the examination, while the patient is lying on his back (supine), he complains of severe pain when his right leg is raised by the clinician.

▶ What is the most likely diagnosis?
▶ What is the anatomical mechanism for this condition?

ANSWER TO CASE 33:

Prolapsed Lumbar Nucleus Pulposus

Summary: A 54-year-old man has lower back pain that radiates down the back of his right leg, is exacerbated by intraabdominal pressure (Valsalva maneuver), and is relieved by rest. He denies trauma. The neurological evaluation is normal, but his pain is elicited by raising a straightened right leg.

- **Most likely diagnosis:** Herniated lumbar disk (prolapsed lumbar nucleus pulposus)

- **Anatomical mechanism for this condition:** Ruptured intervertebral disk impinging on the nerve root as it exits from the vertebral canal

CLINICAL CORRELATION

This patient experiences pain radiating down the back of his leg in the distribution served by the sciatic nerve. Hence, the syndrome is referred to as **sciatica.** The pain is caused by impingement of the nerve roots contributing to the sciatic nerve (L4–S3). He has no history of trauma, and we do not have details of his occupation. Heavy lifting is often an associated factor. The pain is worsened by increased intraabdominal pressure (Valsalva maneuver); thus, coughing and straining often exacerbate the symptoms. The straight leg raising maneuver elicits pain. Because this patient does not have neurological deficits, conservative therapy would include rest, physical therapy, and nonsteroidal anti-inflammatory agents. Most patients improve with this treatment. Lack of improvement, neurological deficits, or history of trauma or malignancy usually necessitates imaging of the spine. MRI is considered to be the most accurate means of examining this region.

APPROACH TO:

The Spine

OBJECTIVES

1. Be able to identify the features of a typical vertebra and the intervertebral joints

2. Be able to label the components of the spinal nerve from spinal roots to primary rami

3. Be able to locate sites where components of the spinal nerve can be compressed

4. Be familiar with the dermatomes and the landmarks of the lower extremity

DEFINITIONS

SUPINE VERSUS PRONE: Supine is the position of lying on one's back, whereas prone is lying on one's stomach.

HERNIATE: To push through a containing membrane or tissue.

SCIATICA: Syndrome caused by irritation to the roots (radiculopathy) of the sciatic nerve.

SYMPHYSIS: A secondary cartilaginous joint, in which two cartilaginous surfaces are held in place by a fibrous connective tissue, such as the intervertebral disk.

DISCUSSION

The **vertebral column** is a series of individual bones that are stacked vertically and held together by ligaments and muscles. There are 32 to 34 vertebrae (7 cervical, 12 thoracic, 5 lumbar, 5 sacral, and 3 to 5 coccygeal). The joints between each vertebra provide flexibility, but the vertebrae are held tightly in place by numerous supporting ligaments that provide strength and stability (Figure 33-1).

The **main features of a typical vertebra are the tubular body** and the **posterior arch** that surrounds and protects the spinal cord. The arch is composed of pedicles that arise from the **vertebral body and lamina that join at the midline.** Each vertebra has **seven processes:** three serve as attachment sites for muscles, and four serve as articular surfaces for adjacent vertebrae. The two **transverse processes** arise from the arch, where the pedicles and laminae meet. One spinous process emerges from the middle of the posterior arch.

Two types of joints support the articulation of adjacent vertebrae. The flat surfaces of the vertebral bodies join through a secondary cartilaginous joint, or symphysis. The bones themselves are separated by **the intervertebral disc,** which has an **outer fibrous layer, the anulus fibrosus, that surrounds a soft inner layer, the nucleus pulposus.** The disc provides support for the joint but also provides flexibility and a cushion against the weight of the upper body. Secondary support is provided by the four articular processes. These processes also emerge from the posterior arch. Two are

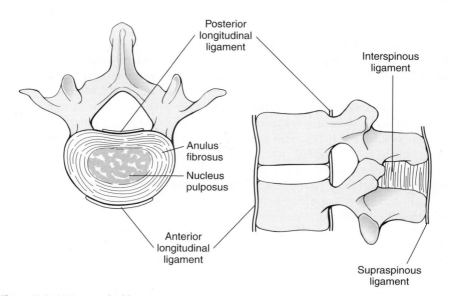

Figure 33-1. Major vertebral ligaments.

directed superiorly and two inferiorly. The superior and inferior processes of adjacent vertebrae join to form a zygapophyseal joint. This synovial joint provides strength with a limited amount of flexibility.

The pedicle and superior articular process together form a notch that is complemented by a second notch formed by the pedicle and inferior process. When two vertebrae are in apposition, the superior and inferior notches form the **intervertebral foramen.** This space is where spinal nerves emerge from the spinal cord to supply peripheral structures.

Peripheral nerve fibers arising from the spinal cord as **anterior (ventral) roots** are primarily **motor**, whereas the **posterior (dorsal) roots are primarily sensory.** These roots join to form the spinal nerve. In the cervical spine, the roots travel laterally to leave the vertebral column. The spinal nerve splits to form two mixed-function branches, a small posterior primary ramus and a larger anterior primary ramus. Nerves emerging from lower levels of the spinal cord course inferiorly before they exit. This is because the **cord itself stops at about vertebral level L1.** Therefore, the roots must travel nearly straight inferiorly before forming the spinal nerves of the lower lumbar, sacral, and coccygeal regions. As these numerous roots stream inferiorly, they form the **cauda equina.**

The symphysis between vertebral bodies is normally very strong because the intervertebral disk is reinforced by **anterior and posterior longitudinal ligaments.** However, in some people, these ligaments weaken, and the intervertebral disk pushes through. If so, the **roots** may be **compressed by the nucleus pulposus through the weakened anulus.** The most common result is stimulation of pain fibers in posterior roots. More serious cases may result in **paresthesia** (area of localized numbness), but rarely is motor function disrupted.

Although the actual site of injury is proximal, the brain perceives the information as coming from the region of the body innervated by the compressed root. Thus, with lumbar herniations, the distribution of this type of pain (radicular pain) tends to follow the **dermatomes of the lower extremity.** These areas progress on the anterior surface from **L1 in the inguinal region** to **L4 at the knee and medial leg** and to **L5 along the lateral leg.** On the posterior surface, **S1 is lateral on the thigh and leg,** and **S2 is medial. S3 through S5 are perianal.** Sensory fibers from a given spinal level spread into adjacent dermatomes. Therefore, in order to achieve complete numbness of a single dermatome, three adjacent spinal nerves must be anesthetized.

In this case, the patient experienced pain when, in the supine position, his straightened leg was raised. This sign indicates that slight mechanical stretching of the sciatic nerve is sufficient to enhance the effect of the herniated disk. Dorsiflexion of the foot exacerbates the pain. In some patients, straightening the contralateral leg may also cause pain in the affected leg, thus confirming radiculopathy.

Radiographic imaging can be used to confirm the herniation. Currently, the best modality is magnetic resonance imaging (**MRI**) because the herniation can be observed directly and MRI is a noninvasive procedure. With the widespread use of MRI, it has become clear that many herniated disks are asymptomatic. An older technique, myelography, is also used on occasion. This technique takes advantage of the fact that the dura mater covers the spinal roots and proximal spinal nerve. Injection of contrast medium into the cerebrospinal fluid (CSF) will infiltrate to the spinal nerves. Therefore, compressed nerve sheaths will not be filled by the dye, and the herniated disc can be observed indirectly.

COMPREHENSION QUESTIONS

33.1 A 34-year-old woman is undergoing cystoscopic examination under spinal anesthesia. As the anesthesiologist places the needle into the subarachnoid space to inject the anesthetic agent, the needle traverses various layers. Which of the following describes the accurate sequence of layers from skin to subarachnoid space?

A. Skin, supraspinous ligament, interspinous ligament, posterior longitudinal ligament, dura mater, subarachnoid space

B. Skin, supraspinous ligament, interspinous ligament, dura mater, subarachnoid space

C. Skin, supraspinous ligament, intertransverse ligament, arachnoid space, subarachnoid space

D. Skin, interspinous ligament, anterior longitudinal ligament, dura mater, subarachnoid space

33.2 A 45-year-old man complains of shooting pain down his right leg that worsens with sitting and coughing. He also has some numbness in the area. The physician tests sensation on the lateral thigh region. Which of the following nerve roots is being tested?

A. L1 and L2

B. L2 and L3

C. L4 and L5

D. S1 and S2

E. S3 and S4

33.3 A 50-year-old diabetic man is having difficulty voiding urine. On examination, he has decreased sensation of the perineal region. Which of the following reflexes is the most likely to be affected?

A. Patellar tendon

B. Achilles tendon

C. Cremaster

D. Anal wink

ANSWERS

33.1 **B.** The sequence of structures is skin, supraspinous ligament, interspinous ligament, dura mater, and subarachnoid space.

33.2 **C.** The lateral thigh is innervated by nerve root L5.

33.3 **D.** The sensory fibers affected are S2 through S4, which innervate the perineal region and supply the afferent limb of the anal wink reflex.

ANATOMY PEARLS

▶ The vertebral column consists of 34 vertebrae: 7 cervical, 12 thoracic, 5 lumbar, 5 sacral, and 3 to 5 coccygeal.

▶ Peripheral nerve fibers emerging from the spinal cord as anterior (ventral) roots are primarily motor, whereas the posterior (dorsal) roots are primarily sensory.

▶ The dermatomes of the lower extremities are L1 in the inguinal region and L4 at the knee and medial leg and L5 along the lateral leg. On the posterior surface, S1 is lateral on the thigh and leg, and S2 is medial. S3 through S5 are perianal.

REFERENCES

Gilroy AM, MacPherson BR. *Atlas of Anatomy*, 2nd ed. New York, NY: Thieme Medical Publishers; 2012:14–15.

Moore KL, Dalley AF, Agur AMR. *Clinically Oriented Anatomy*, 7th ed. Baltimore, MD: Lippincott Williams & Wilkins; 2014:464–465, 474–476.

Netter FH. *Atlas of Human Anatomy*, 6th ed. Philadelphia, PA: Saunders; 2014: plates 155–159.

A 68-year-old man complains of severe burning and stinging pain across the right side of his waist over a period of 2 days. Today, he notes a rash breaking out in the same area. On examination, there is a red rash with blisters starting on his back and curving down and across his right waist region.

▶ What is the most likely diagnosis?
▶ What is the anatomical explanation for this condition?

ANSWER TO CASE 34:

Herpes Zoster

Summary: A 68-year-old man has had severe dysesthesia on his right waist for 2 days and, more recently, an erythematous vesicular (blisterlike) rash.

- **Most likely diagnosis:** Herpes zoster

- **Anatomical explanation for this condition:** Reactivation of the varicella virus and infection of the skin following the dermatomal distribution, which is in this case likely T11 or T12

CLINICAL CORRELATION

This 68-year-old man has clinical symptoms consistent with herpes zoster, also known as "shingles." The chicken pox virus remains latent and may become reactivated years later as a result of illness, stress, or age. The varicella virus is reactivated from the dorsal root ganglia and initially causes a burning pain that follows the distribution of a dermatome, most commonly T3 through L3. Usually 2 to 3 days after the pain, a rash erupts that is erythematous and vesicular, and has a reddish blisterlike appearance. Treatment of this condition may include corticosteroid therapy, which can help to decrease the inflammation and pain. Even after the skin lesions have healed, the patient can have significant pain, called **postherpetic neuralgia**. The pain can persist for months or even years. Treatment of postherpetic neuralgia is difficult, and therapies include topical lidocaine gel, capsaicin cream, anticonvulsant agents, or even nerve blocks.

APPROACH TO:

The Spinal Nerves

OBJECTIVES

1. Be able to draw the components of a spinal nerve

2. Be able to draw the dermatomes of the thorax and abdomen

DEFINITIONS

DYSESTHESIA: An abnormality of somatic sensation; for example, diminished sensation approaching numbness, an uncomfortable or painful response to a normal stimulus (as in this case) or a perceived sensation in the absence of stimulation

ERYTHEMATOUS VESICULAR RASH: Skin disorder characterized by redness and pustules

DISCUSSION

Peripheral nerve fibers arise from the spinal cord as **anterior (ventral) roots**, which are primarily **motor**, and **posterior (dorsal) roots, which are primarily sensory.** These roots join to form the spinal nerve. The spinal nerve splits to form two mixed function branches, a **small posterior primary ramus and a larger anterior primary ramus.** In the abdomen, the posterior primary ramus innervates the intrinsic muscles of the back and the overlying skin. The anterior primary ramus projects anteriorly and inferiorly to innervate the muscles of the abdominal wall and the overlying skin (Figure 34-1). For a description of the peripheral distribution of spinal nerves, see Case 33.

After infecting the skin, the varicella virus is transported within the axons of sensory neurons back to the cell bodies, which are located in the **posterior root ganglia.** The virus periodically reactivates and is transported back out along the distribution of the spinal nerve that is carrying the sensory axons. Thus skin eruptions occur along the dermatomal distribution of the spinal nerve. In the thorax, the main dermatomal landmarks are the clavicle (L5) and nipple (T4). In the abdomen, the major dermatomal landmarks are the **xiphoid process of the sternum (T7), umbilicus (T10), and inguinal/suprapubic region (L1) (Figure 34-2).**

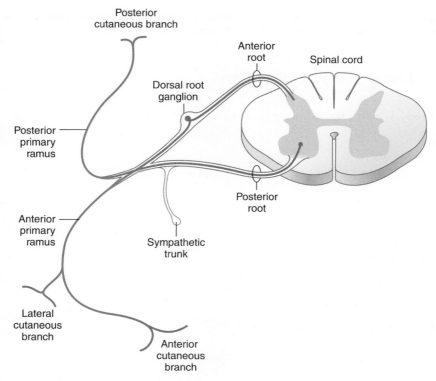

Figure 34-1. Components of a typical spinal nerve.

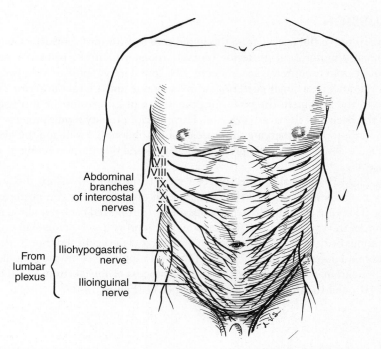

Figure 34-2. Distribution of cutaneous nerves over the abdomen. (*Reproduced, with permission, from Lindner HH. Clinical Anatomy. East Norwalk, CT: Appleton & Lange, 1989:302.*)

COMPREHENSION QUESTIONS

34.1–34.4 Match the following nerve roots (A–F) to their locations on the body.

A. C5

B. C7

C. T1

D. T4

E. T7

F. T10

34.1 Umbilicus

34.2 Nipple

34.3 Xiphoid process

34.4 Clavicle

ANSWERS

34.1 **F.** T10 innervates the skin around the umbilicus.

34.2 **D.** T4 innervates the nipple area.

34.3 **E.** T7 innervates the xiphoid process of the sternum.

34.4 **A.** C5 innervates the skin over the clavicle.

ANATOMY PEARLS

▶ Peripheral nerve fibers arise from the spinal cord. Anterior (ventral) roots are primarily motor, whereas posterior (dorsal) roots are primarily sensory.

▶ In the abdomen, the major dermatomal landmarks are the xiphoid process of the sternum (T7), umbilicus (T10), and inguinal and suprapubic regions (L1).

▶ In the thorax, the major dermatome landmarks are the clavicle (C5) and the nipple (T4).

REFERENCES

Gilroy AM, MacPherson BR, Ross LM. *Atlas of Anatomy*, 2nd ed. New York, NY: Thieme Medical Publishers; 2012:42–43, 66–68.

Moore KL, Dalley AF, Agur AMR. *Clinically Oriented Anatomy*, 7th ed. Baltimore, MD: Lippincott Williams & Wilkins; 2014:193–195, 496–498.

Netter FH. *Atlas of Human Anatomy*, 6th ed. Philadelphia, PA: Saunders; 2014: plates 162, 174, 253–254.

A 7-year-old boy is brought into the emergency room for severe headache, nausea, and fever. His parent states that the patient had been in good health until 2 days previously. Bright lights seem to bother him. On examination, he appears lethargic and ill. His temperature is 102°F. Movement of the neck seems to cause some pain. The heart and lung examinations are normal. The patient refuses to flex his head to enable his chin to touch his chest because the effort is too painful.

▶ What is the most likely diagnosis?
▶ What is the most likely anatomical mechanism for this condition?

ANSWER TO CASE 35:

Meningitis

Summary: A 7-year-old boy has a 2-day history of severe headache, nausea, fever, and photophobia. He appears lethargic and ill. His temperature is 102°F, and he has some nuchal rigidity.

- **Most likely diagnosis:** Meningitis
- **Most likely anatomical mechanism:** Migration of bacteria through the nasopharynx or through the choroid plexus

CLINICAL CORRELATION

This young child has a 2-day history of fever, headache, nausea, and photophobia. His general appearance suggests sepsis, and the nuchal rigidity suggests meningitis. These symptoms are caused by cerebral inflammation, ischemia, and edema. Increased intracranial pressure may cause lethargy and even seizures. The most common causative organisms are *Streptococcus pneumoniae* and *Neisseria meningitidis.* Previously, *Haemophilus influenzae* was the most commonly isolated organism; however, with the advent of the *H. influenzae* vaccine, this pathogen has been less of a factor. The diagnosis is made by lumbar puncture. Positive findings would be leukocytes (in particular neutrophils) isolated from the cerebrospinal fluid (CSF), and grampositive organisms. A diffuse erythematous maculopapular rash that becomes petechial is suggestive of meningococcus. Rapid initiation of empiric antibiotic therapy is critical; the medication is aimed at the most common causative organisms and must penetrate through the blood-brain barrier.

APPROACH TO:

Meninges and CSF

OBJECTIVES

1. Be able to identify the meningeal layers
2. Be able to draw the flow of CSF from the choroid plexus to the subarachnoid space
3. Be able to identify sites where infection may spread into the cranial cavity
4. Be able to describe the innervation of the meninges and the pathogenesis of headache

DEFINITIONS

ISCHEMIA: Decrease in blood flow to a tissue, generally due to blockage of the nutrient arteries

EDEMA: Swelling due to accumulation of water in tissue

PETECHIAE: Tiny spots in the skin generally due to broken capillaries

CHOROID PLEXUS: A tissue lining the ventricles of the brain that produces the CSF that fills the ventricles and the subarachnoid space

MENINGITIS: A very serious infection involving the meninges, which may be caused by viruses or bacteria and may lead to long-term consequences or death

DISCUSSION

Within the cranial cavity, the **brain is protected by three meningeal layers.** The **dura mater,** a **thick fibrous membrane,** is the most **superficial.** Apposed to the deep surface of the dura mater is the **arachnoid mater,** which is a delicate, thin membrane that is nearly transparent. The **pia mater** is the thinnest layer, and it is **directly apposed to the surface of the brain.** Three spaces relate to the three layers. The **epidural** space lies between the periosteum of the calvaria and the dura mater. Normally, the dura is closely apposed to the bone, so this is a potential space that can be expanded by blood or pus. Similarly, the arachnoid mater is closely apposed to the dura mater. The subdural space between the two layers is also a potential space. The **subarachnoid space** lies between the **arachnoid and pia mater.** This space is normally filled with CSF, which is the extracellular fluid of the central nervous system. CSF pooled in the subarachnoid space also serves a protective function by helping to insulate the brain and spinal cord from mechanical shocks.

CSF in the subarachnoid space is **produced** by the **choroid plexus,** in the ependyma of the lateral, third, and fourth ventricles. CSF produced in the lateral ventricles flows through the **interventricular foramina of Monro** into the **third ventricle.** The **cerebral aqueduct of Sylvius** then conducts CSF into the **fourth ventricle.** From there, fluid flows through the **foramina of Magendie and Luschka** into the subarachnoid space, where it surrounds the brain and spinal cord. The circulatory path ends at the arachnoid granulations, where CSF is resorbed back into the venous system. Most arachnoid granulations are found lining the large venous sinuses, but arachnoid villi may also be present at the roots of spinal nerves.

Meningitis is an **inflammation of the meninges,** but in practice the term refers to **infections of the pia and arachnoid layers,** usually involving the CSF (Figure 35-1). Infections reach the meninges by several routes. Most infections seem to be transferred through the vasculature (hematogenous transmission). On the arterial side, bacteria can infiltrate through the choroid plexus into the CSF. On the venous side, there are several routes from the face into the cranium. Normally, the veins drain superficially and inferiorly through the **pterygoid venous plexus** and **facial and retromandibular veins.** However, there are also anastomoses with the **superior and inferior ophthalmic veins.** These veins carry blood from the orbit into the cavernous sinus, which is in the middle cranial fossa. Because veins in the face have no valves, some infections can reverse the normal flow of blood so that **pathogens are carried into the cavernous sinus.** They can then infiltrate through the walls of the sinus into the CSF. A second route is through the nasopharynx. Mucosal infections can track through the **cribriform plate** into the anterior cranial fossa.

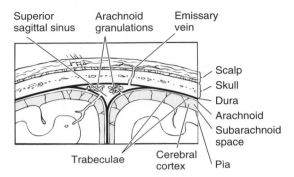

Figure 35-1. Cross-sectional view of meninges. (*Reproduced, with permission, from Waxman SG. Lange's Clinical Neuroanatomy, 25th ed. New York: McGraw-Hill, 2003:158.*)

The **severe headaches associated with meningitis are due to increased intracranial pressure**, which **stretches the dura mater** and stimulates pain fibers from **cranial nerve V3** (mandibular branch of the trigeminal nerve) traveling with branches of **the middle meningeal artery.**

COMPREHENSION QUESTIONS

35.1–35.4 Match the following anatomical spaces (A–D) to the descriptions of location.

    A. Epidural

    B. Subdural

    C. Subarachnoid

    D. Intraarachnoid

35.1 Between the dura mater and the calvaria

35.2 Between the arachnoid and pia maters

35.3 Between the dura and arachnoid maters

35.4 A 2-month-old baby is noted to have macrocephaly (large head) and developmental delay. On ultrasound, the baby has significant hydrocephalus. The pediatrician is suspicious of congenital stenosis of the aqueduct of Sylvius. Which of the following are the most likely findings in this infant?

	Right Lateral Ventricle	Left Lateral Ventricle	Third Ventricle	Fourth Ventricle
A.	Dilated	Normal	Normal	Normal
B.	Dilated	Dilated	Normal	Normal
C.	Dilated	Dilated	Dilated	Normal
D.	Dilated	Dilated	Dilated	Dilated
E.	Normal	Normal	Dilated	Dilated
F.	Normal	Normal	Normal	Dilated

ANSWERS

35.1 **A.** The epidural space is between the fibrous dura mater and the periosteum of the calvaria.

35.2 **C.** The subarachnoid space is between the arachnoid and the pia maters.

35.3 **B.** The subdural space is between the dura and the arachnoid maters.

35.4 **C.** The aqueduct of Sylvius is between the third and fourth ventricles; thus, dilation of the lateral ventricles and the third ventricle is seen with aqueductal stenosis.

ANATOMY PEARLS

▶ The three meningeal layers that protect the brain are the dura mater (close to the skull), the arachnoid mater, and the pia mater (adherent to the brain).

▶ The epidural space is located between the fibrous dura mater and the periosteum of the calvaria, the subdural space is between the dura and pia maters, and the subarachnoid space is between the arachnoid and pia maters.

▶ CSF is produced by the choroid plexus, which is located in the lateral and fourth ventricles.

▶ The lateral ventricles connect to the third ventricle through the interventricular foramina of Monro. CSF flows through the cerebral aqueduct of Sylvius into the fourth ventricle and then flows through the foramina of Magendie and Luschka into the subarachnoid space.

▶ Meningitis is an infection of the pia and arachnoid layers, usually involving the CSF.

REFERENCES

Gilroy AM, MacPherson BR, Ross LM. *Atlas of Anatomy,* 2nd ed. New York, NY: Thieme Medical Publishers; 2012:481, 524–526.

Moore KL, Dalley AF, Agur AMR. *Clinically Oriented Anatomy,* 7th ed. Baltimore, MD: Lippincott Williams & Wilkins; 2014:498–501, 865–867, 878–881.

Netter FH. *Atlas of Human Anatomy,* 6th ed. Philadelphia, PA: Saunders; 2014: plates 101, 103–105.

A 35-year-old woman complains of a 2-month history of hoarseness and some choking while drinking liquids. She denies viral illnesses. She underwent surgery for a cold nodule of the thyroid gland 9 weeks ago. Her only medication is acetaminophen with codeine.

▶ What is the most likely diagnosis?
▶ What is the anatomical explanation for her symptoms?

ANSWER TO CASE 36:

Recurrent Laryngeal Nerve Injury

Summary: A 35-year-old woman has a 2-month history of voice hoarseness and choking after undergoing surgery for a cold (nonfunctioning) thyroid nodule

- **Most likely diagnosis:** Injury to the recurrent laryngeal nerve
- **Anatomical explanation for her symptoms:** Vocal cord paralysis

CLINICAL CORRELATION

This woman underwent surgery for a thyroid nodule. A **cold nodule** is defined as a mass that does not take up radioactive (i.e., "hot") iodine isotope. Surgery of the thyroid gland can sometimes injure the recurrent laryngeal nerve, which runs through the posterior superior suspensory ligament of the thyroid gland. The recurrent laryngeal nerve provides motor innervation to the larynx and sensory innervation to the laryngeal mucosa. A traction injury or inadvertent severing of the nerve leads to vocal cord paralysis. With injury to just one nerve, the vocal cord on the same side bows into a paramedian position instead of closing straight to the midline, leading to hoarseness. When drinking, the patient may choke if the liquid is aspirated into the trachea. When vocal cord function does not return after 6 months to 1 year, then injection of the affected vocal cord with Teflon can be helpful.

There are four small parathyroid glands within the thyroid tissue, usually two in the left lobe and two in the right lobe of the thyroid gland. These tiny parathyroid glands secrete parathyroid hormone to maintain calcium balance. Inadvertent injury due to excision of the parathyroid glands can lead to hypocalcemia, manifested by fatigue, dyspnea (shortness of breath), brittle skin and nails, tetanic muscle contractions, seizures, or difficulty swallowing.

APPROACH TO:

The Neck: Thyroid Gland

OBJECTIVES

1. Be able to identify the parts of the thyroid gland
2. Be able to draw branches of the arteries and veins that supply the thyroid gland
3. Be able to identify the main features of the larynx and list features that assist in respiration (phonation) or protect the laryngeal inlet during swallowing
4. Be able to identify the course of the different branches of the vagus nerve cranial nerve (CN) X that innervate the larynx
5. Be able to describe the consequences of injury to the recurrent laryngeal nerve and contrast with the consequences of injury to the superior laryngeal nerve

DEFINITIONS

COLD NODULE: A region of the thyroid gland that does not take up hot iodine radioisotope (as visualized with thyroid scintigraphy) because the tissue does not contain follicular thyroid cells

ECTOPIC: Tissue that resides in an unexpected or abnormal location

ASPIRATE: To suck food or liquid into the bronchial tree of the lungs, possibly resulting in inflammation or pneumonia

DISCUSSION

The **thyroid gland** is located at the **base of the neck.** It consists of **left and right lobes** connected by a **narrow isthmus.** During development, the **gland forms at the base of the tongue at the foramen cecum** and **descends into the neck** along the **thyroglossal duct,** reaching its final position **inferior to the cricoid cartilage** (vertebral levels C5 through T1). Occasionally, ectopic thyroid tissue will deposit along the duct. This sometimes manifests as a pyramidal lobe arising from the midline along the remnants of the duct.

As an endocrine gland, the thyroid receives a rich vascular supply. The **superior thyroid artery** is the **first anterior branch of the external carotid artery.** It descends laterally to the hyoid bone, giving off the **superior laryngeal artery,** which pierces the thyrohyoid membrane. The **superior thyroid artery** continues toward the gland lateral to the thyroid and cricoid cartilages. It crosses along the superior border of the thyroid and usually anastomoses with the contralateral superior thyroid artery. The **inferior thyroid artery** is a branch of the **thyrocervical trunk,** which arises from the first part of the subclavian artery. The artery ascends, giving off an ascending cervical artery, and then curves inferiorly to enter the thyroid gland from the posterior surface. There are many **anastomoses between branches of the superior and inferior thyroid arteries.** Rarely, an artery arising directly from the brachiocephalic trunk or the aortic arch, called the **thyroidea ima artery,** will ascend to supply the thyroid. The gland is drained by three pairs of veins. The superior and middle thyroid veins drain to the internal jugulars, and the inferior thyroid veins drain to the brachiocephalics.

The **thyroid lies anterior to the trachea (Figure 36-1),** a hollow tube that conducts air to the lungs. It forms from the inferior pharynx, anterior to the esophagus. The wall of the trachea is supported by a series of cartilaginous rings. All of the unnamed rings have a **C shape,** leaving the posterior wall flexible to accommodate expansion of the esophagus during swallowing. Superior to the thyroid gland are the **cricoid and thyroid cartilages.** These are specialized structures that protect the underlying structures of the larynx.

The structures of the **larynx** serve two functions: to modulate expelled air to make sounds used in the **production of speech** and to **protect the airway** from food and drink passing to the esophagus. The larynx consists of three single cartilages (**epiglottis, thyroid,** and **cricoid**), and three paired cartilages (**arytenoid, corniculate,** and **cuneiform**), for a total of nine. The **thyroid cartilage** resembles a partially open book, with the union of its two plates forming a **laryngeal prominence** (Adam's apple) anterosuperiorly. **Inferior to the thyroid cartilage is the cricoid cartilage**

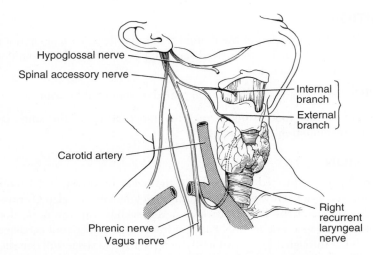

Figure 36-1. Nerves related to the thyroid gland. (*Reproduced, with permission, from Lindner HH. Clinical Anatomy. East Norwalk, CT: Appleton & Lange, 1989:138.*)

which resembles a signet ring. It's large lamina is posterior. Posterior to the thyroid cartilage is the **epiglottis**, a cartilaginous structure that is bound to the thyroid cartilage by the thyroepiglottic ligament. The **arytenoid cartilages** rest on the superior margin of the cricoid cartilage and are held in place by capsules that surround the cricoarytenoid joint. The epiglottis attaches to the arytenoid cartilages through the quadrangular membrane. The free superior border forms the aryepiglottic fold, and the free inferior border forms the vestibular ligament (false vocal fold). The other major structure of the larynx is the **conus elasticus**, another broad ligament inferior to the quadrangular membrane. This ligament is the fusion of the lateral and median cricothyroid ligaments. The free superior border also attaches to the arytenoid cartilage and forms the **vocal ligament (true vocal fold).** The space between the two vocal folds is the **rima glottidis.** When the rima glottidis is wide (i.e., the folds are abducted), maximal air flow is permitted through the trachea. When the rima glottidis is closed (i.e., the folds are adducted), no air flows. When the **rima glottidis is narrow, the expelled air will vibrate the vocal folds and produce a sound.**

The intrinsic musculature of the larynx is devoted mostly to fine-motor control of the vocal folds to modulate pitch and intonation during speech. Perhaps the most important muscles are the **posterior cricoarytenoids**, which are the only muscles used to abduct the vocal folds and are **necessary to widen the rima glottidis for breathing.** All of the other muscles function to adduct the rima glottidis or modulate the tension of the vocal chords. The **lateral cricoarytenoids** adduct the vocal folds. The transverse and oblique arytenoid muscles bring the two arytenoid cartilages together, which indirectly act to close the posterior portion of the rima glottidis. The cricothyroids lengthen and tighten the vocal fold, whereas the thyroarytenoid relaxes it. The vocalis muscle runs under the vocal fold and produces local modulations in tightness (e.g., relaxing posteriorly while tightening anteriorly; Figure 36-2).

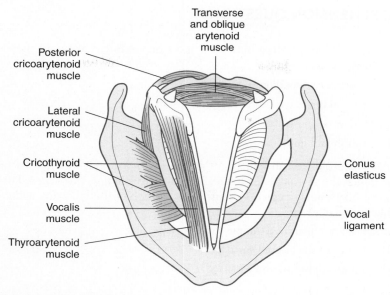

Figure 36-2. Larynx showing vocal cords.

Several structures protect the trachea from food or liquid traveling to the esophagus. The first of these is the **epiglottis**, which deflects food laterally around the quadrangular membrane to the piriform recess and into the esophagus. The epiglottis itself does not apply sufficient force to completely close off the laryngeal inlet. During swallowing, the **suprahyoid muscles** contract and, through the thyrohyoid membrane, lift the larynx up against the epiglottis. The **infrahyoid muscles** attached to the external face of the thyroid cartilage help to return the larynx to its resting position.

Most of these intrinsic laryngeal muscles are innervated by the **recurrent laryngeal nerve**, a branch of the **vagus nerve (CN X).** The only exception is the cricothyroid, which is innervated by the external branch of the superior laryngeal nerve, also a branch of the vagus. Thus, **damage to the superior laryngeal nerve** will affect voice quality, particularly the ability to reach high tones. More significantly, damage to the **recurrent laryngeal nerve will impair** the ability to abduct the vocal folds, possibly leading to respiratory distress if the injury is bilateral. Unilateral damage to the recurrent laryngeal nerve will result in inability to tightly adduct the two vocal folds, resulting in **hoarseness.** In addition, the **protective function of the rima glottidis may be lost,** and food or liquid that does not go down the esophagus may flow into the trachea and cause a **choking response.** In extreme cases, **aspiration pneumonia** may result.

Sensory innervation of the larynx is also mediated by the vagus nerve. In the **supraglottic region** (above the vocal fold), the mucosa is innervated by the internal branch of the superior laryngeal nerve. In the **infraglottic region** (below the fold), the mucosa is innervated by the **recurrent laryngeal nerve**. Thus, damage to the superior and recurrent laryngeal nerves may also have deficits in reflex behaviors that depend on sensory input from the larynx.

COMPREHENSION QUESTIONS

36.1 Which of the following muscles is most important to allow air movement through the larynx?

A. Posterior cricoarytenoids

B. Lateral cricoarytenoids

C. Cricothyroid muscle

D. Infrahyoid muscles

36.2 A 33-year-old woman underwent partial thyroidectomy for hyperthyroidism in which the thyroid failed to take up radioactive iodine. She is noted to have some hoarseness of voice 1 month later. Which of the following is the most likely explanation?

A. Endotracheal tube trauma to the vocal cords

B. Injury to the cricoid cartilage

C. Injury to the thyroid cartilage

D. Injury to the recurrent laryngeal nerve

36.3 A 15-year-old boy is eating a fish dinner and inadvertently has a bone "caught in his throat." He complains of significant pain above the vocal cords. Which of the following nerves is responsible for carrying the sensation for this pain?

A. Superior laryngeal nerve

B. Recurrent laryngeal nerve

C. Spinal accessory nerve

D. Hypoglossal nerve

36.4 A 25-year-old woman underwent surgery for a thyroid nodule. Two months later, she complains of dryness of skin and muscle spasms. Which of the following is the most likely explanation?

A. Low serum magnesium

B. Low serum calcium

C. Low serum potassium

D. Low serum sodium

E. Low serum glucose

ANSWERS

36.1 **A.** The posterior cricoarytenoid muscles are the only muscles that abduct the vocal folds and are necessary to widen the rima glottidis for breathing.

36.2   **D.** Injury to the recurrent laryngeal nerve is common during thyroid surgery and may lead to the inability to tightly adduct the two vocal folds, resulting in hoarseness. In addition, the protective function of the rima glottidis may be lost, and food or liquid that does not go down the esophagus may flow into the trachea and cause a choking response.

36.3   **A.** The laryngeal mucosa above the vocal cords is innervated by the superior laryngeal nerve, whereas mucosa below the vocal cords is innervated by the recurrent laryngeal nerve.

36.4   **B.** This patient likely has hypocalcemia due to excision of the parathyroid glands.

ANATOMY PEARLS

► The right recurrent nerve is located more laterally than the left recurrent nerve because of the course of the right subclavian artery.

► The posterior cricoarytenoids are the only muscles to abduct the vocal folds and are necessary to widen the rima glottidis for breathing.

► Most of the intrinsic laryngeal muscles are innervated by the recurrent laryngeal nerve, a branch of the vagus nerve (CN X).

► Bilateral injury to the recurrent laryngeal nerves may lead to respiratory distress, whereas unilateral injury results in hoarseness.

► Injury to the recurrent laryngeal nerve may affect the protective function of the rima glottidis, increasing the opportunity for a choking response.

REFERENCES

Gilroy AM, MacPherson BR, Ross LM. *Atlas of Anatomy,* 2nd ed. New York, NY: Thieme Medical Publishers; 2012:106, 600, 603, 607.

Moore KL, Dalley AF, Agur AMR. *Clinically Oriented Anatomy,* 7th ed. Baltimore, MD: Lippincott Williams & Wilkins; 2014:1016–1017, 1030, 1076.

Netter FH. *Atlas of Human Anatomy,* 6th ed. Philadelphia, PA: Saunders; 2014: plates 76–78.

A 59-year-old man complains of numbness of his right arm and slurred speech for a 4-h duration. On examination, he has a blood pressure of 150/90 mmHg and a normal body temperature. His heart has a regular rate and rhythm. Auscultation of the neck reveals a blowing sound bilaterally.

▶ What is the most likely diagnosis?
▶ What is the most likely anatomical mechanism for this condition?

ANSWER TO CASE 37:

Carotid Insufficiency

Summary: A 59-year-old hypertensive man complains of a 4-h history of numbness of his right arm and slurred speech. He has carotid bruits bilaterally.

- **Most likely diagnosis:** Transient ischemic attack

- **Most likely anatomical mechanism:** Left carotid insufficiency leading to ischemia of the left cerebral hemisphere

CLINICAL CORRELATION

This 59-year-old man has a 4-h history of right arm numbness and slurred speech. This suggests ischemia of the left cerebral hemisphere including the speech area. If the deficits were to resolve before 24 h, it would be called a **transient ischemic attack**. If the deficits were to continue beyond 24 h, it would be called a **cerebrovascular accident**, or **stroke**. The two major types of strokes are ischemic and hemorrhagic. Differentiating between the two is important because fibrinolytic therapy (medication that dissolves blood clots) would be contraindicated with hemorrhagic strokes. Ischemic strokes can be caused by atherosclerosis or emboli. In this patient, the bruits identified on the carotid arteries are likely due to increased rate and turbulence of blood flow through the stenotic vessels. Immediate management of this patient would include administration of an antiplatelet medication such as aspirin and/or clopidogrel. An emergency computer tomographic (CT) scan of the head can help to differentiate between ischemic and hemorrhagic stroke, and fibrinolytic therapy can be considered if a cerebral hemorrhage is not found. After stabilization of the patient, carotid endarterectomy surgery may be indicated.

APPROACH TO:

The Neck: Vasculature

OBJECTIVES

1. Be able to review the somatotopic organization of sensory and motor regions in the brain

2. Be able to list the branches of the common carotid artery and the vascular supply to the brain and identify the sites most susceptible to formation of atherosclerotic plaques

3. Be able to describe structures in the anterior triangle of the neck that relate to the carotid artery and sheath

DEFINITIONS

AUSCULTATION: Procedure of listening to the body during the physical examination, generally through a stethoscope

BRUITS: An abnormal sound heard through the stethoscope, generally a "whoosh"

SOMATOTOPIC: An orderly mapping of the body surface onto an internal organ, usually a region of the brain

ANSA CERVICALIS: A loop formed superficially in the carotid sheath by branches of cervical spinal nerves innervating the strap muscles

DISCUSSION

The major structural features of the **brain** include the **cerebrum and the cerebellum.** The cerebrum is involved in the major functions of **sensory perception, motor control**, and the associational processing that integrates the two. The cerebellum is primarily involved in **motor control.** The surface or cortex of the cerebrum is folded into a number of ridges (**gyri**) separated by valleys (**sulci**) of different depths. The brain is divided into **lobes** named for the overlying cranial bones: **frontal, temporal, parietal, and occipital.** The central sulcus separates the frontal from the parietal lobes. The **precentral gyrus** controls **voluntary motion**, whereas the **postcentral gyrus is the site of somatosensory perception.**

The sensory and motor areas are arranged according to a somatotopic organization. The **lower extremity** is represented **medially along the gyrus**; the **upper extremity, more laterally**; and the **head and neck, most laterally.** The **tracts** going to and from the sensory and motor areas **cross in the lower brain and spinal cord to control the opposite side of the body.** Another important region is the **motor speech area (Broca area)**, which is a **small gyrus in the anterior parietal cortex of the left brain**, called the **operculum**, just superior to the temporal lobe. These basic organizational features are important because they help physicians identify the region of the brain damaged by a stroke or hemorrhage. Thus, numbness or paralysis of the right upper extremity indicates **damage on the left side of the brain**, which will frequently involve the motor speech area.

The **blood supply to the brain** is from the **common carotid and the vertebral arteries** (Figure 37-1). The two ascend separately through the neck. The **vertebral artery** ascends through the **transverse foramina of the cervical vertebrae** without giving off any major branches. It then curves medially to ascend through the foramen magnum. The paired **vertebral arteries fuse to form the basilar artery.** The common carotid bifurcates at about the level of the hyoid bone (vertebrae C3 and C4). The external carotid ascends to provide branches to structures outside the cranium. The **internal carotid** ascends without major branches to enter the cranium through the carotid canal. After a relatively tortuous course through the sphenoid bone and the cavernous sinus, the internal carotid emerges into the **middle cranial fossa.** In the region of the **sella turcica** and surrounding the pituitary stalk, the vertebral and carotid circulations anastomose through a complex structure called **the circle of Willis**, which is formed posteriorly by the bifurcation of the basilar into left and right posterior cerebral arteries. It forms anteriorly with the internal

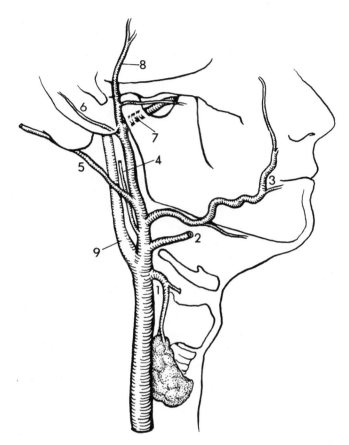

Figure 37-1. Branches of the external carotid artery: 1 = superior thyroid, 2 = lingual, 3 = facial, 4 = ascending pharyngeal, 5 = occipital, 6 = posterior auricular, 7 = maxillary, 8 = superficial temporal, 9 = internal carotid. (*Reproduced, with permission, from the University of Texas Health Science Center, Houston Medical School.*)

carotids, which bifurcate into the anterior and middle cerebral arteries. The left and right anterior cerebrals anastomose through the anterior communicating branch. The middle cerebral and posterior cerebral arteries anastomose through the posterior communicating branch.

The carotids are susceptible to occlusion as a result of atherosclerotic disease. The most common sites of occlusion are bifurcation of the internal and external carotids and bifurcation of the anterior and middle cerebral arteries. Occasionally, small pieces of an atherosclerotic plaque will break off (**embolize**) and obstruct a smaller artery. The **middle cerebral artery and its branches are most commonly** affected by this process. This artery ascends through the Sylvian fissure along the superior border of the temporal lobe. A major branch runs lateral to medial in the central sulcus. As a consequence, many strokes produce deficits in movement and sensation of the face and upper extremities and in language. More severe damage may also affect the lower extremity.

The carotid arteries are palpated in the anterior triangle of the neck. The **sternocleidomastoid muscle (SCM)** **separates the superficial neck into anterior and posterior triangles.** The **anterior triangle is further divided by the omohyoid and digastric muscles into four additional triangles: submental, submandibular, muscular, and carotid.** The common carotid artery ascends in the neck deep to the contents of the muscular triangle. These muscles are the sternothyroid, sternohyoid, and superior belly of omohyoid. It then goes through the carotid triangle, which is bordered by the superior belly of omohyoid, posterior belly of digastric, and sternocleidomastoid muscles. The **common carotid bifurcates** at the **level of the hyoid bone (C3)**, and the internal carotid continues posteriorly. The carotids are contained within the **carotid sheath.** This fascial membrane originates from the other three layers of deep fascia that are in the neck: superficial layer of deep fascia, prevertebral fascia, and pretracheal/buccopharyngeal fascia. Also contained in the carotid sheath are the internal jugular vein and the vagus nerve (CN X). The vein is larger than the artery and lies more superficially. Several other cranial nerves have a structural relation to the carotid sheath. These include the glossopharyngeal (CN IX) and spinal accessory (CN XI) nerves, which exit with the vagus through the jugular foramen. The sinus branch of the glossopharyngeal nerve courses within the sheath to innervate the carotid body and sinus. The hypoglossal nerve (CN XII) passes deep to the carotid sheath as it projects anteriorly into the submandibular triangle. The sympathetic trunk lies deep to the carotid sheath on the surface of the prevertebral muscles. The superior and inferior roots of the ansa cervicalis, from spinal nerves C2 through C4, typically form the loop within the anterior surface of the carotid sheath before giving off branches to the infrahyoid muscles.

COMPREHENSION QUESTIONS

37.1 A 47-year-old man complains of right arm weakness and difficulty speaking (expressive aphasia). Which of the following arteries is most likely affected?

  A. Vertebral

  B. Posterior cerebral

  C. Middle cerebral

  D. Anterior cerebral

37.2 While performing a carotid endarterectomy in a 55-year-old man who has carotid artery occlusion, and on approach to the internal carotid artery, a surgeon severs a nerve embedded in the carotid sheath. Which nerve was severed?

  A. Superior laryngeal

  B. Vagus

  C. Sympathetic trunk

  D. Ansa cervicalis

  E. Recurrent laryngeal

37.3 A 44-year-old man falls from a tree and develops a severe scalp hematoma. The superficial temporal artery continues to bleed internally because the man takes warfarin sodium (Coumadin) for an artificial heart valve. Which of the following arteries may be ligated to control the bleeding?

A. Internal carotid
B. External carotid
C. Occipital
D. Maxillary

ANSWERS

37.1 **C.** The middle cerebral artery supplies the temporal and parietal regions that contain the Broca area (the speech center).

37.2 **D.** The ansa cervicalis is a branch of the cervical plexus that innervates the infrahyoid strap muscles. The superior root generally descends within the carotid sheath superficially to the internal jugular vein. Therefore, this nerve is at risk during surgical approaches to the internal carotid artery.

37.3 **B.** The external carotid artery divides into two branches: the maxillary artery and the superficial temporal artery.

ANATOMY PEARLS

▶ The brain is divided into lobes named for the overlying cranial bones: frontal, temporal, parietal, and occipital.

▶ The blood supply to the brain is from the common carotid and the vertebral arteries

▶ The carotid arteries are palpated in the anterior triangle of the neck.

▶ The vertebral and carotid circulations anastomose through a complex structure called the circle of Willis.

REFERENCES

Gilroy AM, MacPherson BR, Ross LM. *Atlas of Anatomy.* 2nd ed. New York, NY: Thieme Medical Publishers, 2012:516–517, 636–637.

Moore KL, Dalley AF, Agur AMR. *Clinically Oriented Anatomy,* 7th ed. Baltimore, MD: Lippincott Williams & Wilkins; 2014:882–885, 887–888, 1001.

Netter FH. *Atlas of Human Anatomy,* 6th ed. Philadelphia, PA: Saunders; 2014: plates 106, 137, 140, 142–143.

A 3-month-old girl is noted by the pediatrician to have a stiff neck for a 2-month duration. The mother states that the neck seems to be pulled to the right. On examination, the baby's right ear is tilted toward her right side, but her face is turned toward the left. Palpation of the neck reveals a nontender mass of the right anterior neck region.

▶ What is the most likely diagnosis?
▶ What is the anatomical structure affected?

ANSWER TO CASE 38:

Torticollis

Summary: A 3-month-old girl's head seems to be flexed to the right and rotated to the left. Palpation of the neck reveals a nontender mass of the right anterior neck region.

- **Most likely diagnosis:** Torticollis
- **Anatomical structure affected:** Sternocleidomastoid muscle (SCM)

CLINICAL CORRELATION

Torticollis is a deformity usually observed in children as lateral flexion and rotation of the head and neck. Congenital torticollis has an incidence of 3 to 5 per 1000 births. It is thought to be due to a fibrosis of the SCM that develops during infancy and causes shortening of the muscle. A mass can be palpated (about the size of an olive) about 66 percent of the time at the SCM. The etiology is unclear, although it may be associated with breech babies or difficult deliveries. As a result, the baby's head is flexed laterally toward the affected side and rotated contralaterally. Facial asymmetry may be noted. Physical therapy can help in most cases, and surgery is rarely needed.

APPROACH TO:

The Triangle Neck: Anterior

OBJECTIVES

1. Be able to identify surface landmarks of the anterior neck
2. Be able to describe the actions of the SCM

DEFINITIONS

FIBROSIS: Abnormal growth of fibrous connective tissue in response to trauma or infection.

ANTERIOR NECK: Portion of the neck anterior to the trapezius muscle. Two triangles are found within the anterior neck; the anterior triangle, which contains structures anterior to the SCM, and the posterior triangle, which contains structures posterior to the SCM.

DISCUSSION

An important surface landmark on the anterior surface of the neck is the **SCM**, which divides the anterior neck into **anterior and posterior triangles.** The superior head of the muscle attaches to the mastoid process of the temporal bone. Inferiorly,

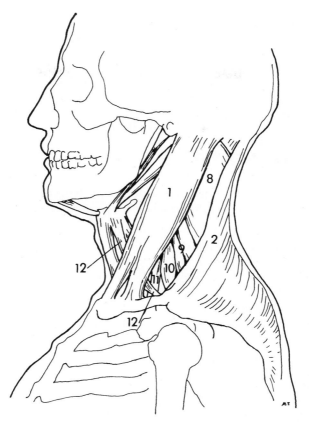

Figure 38-1. The sternocleidomastoid muscle and anterior neck: 1 = sternocleidomastoid muscle, 2 = trapezius muscle, 8 = splenius capitis, 9 = levator scapulae, 10 = middle scalene, 11 = anterior scalene, 12 = omohyoid muscle. (*Reproduced, with permission, from the University of Texas Health Science Center, Houston Medical School.*)

the muscle splits to attach separately to the manubrium of the sternum and the clavicle (Figure 38-1).

Contraction of the SCM has two sequelae: **rotation of the head to the opposite side**—thus, contraction of the right SCM will rotate the nose toward the left; and **lateral flexion**—constant contraction of a single SCM will frequently result in lateral flexion to the affected side and rotation to the opposite side, sometimes called "wry neck." Simultaneous contractions of both SCM muscles may contribute to straight flexion of the neck because the rotational movements cancel each other. However, this is not a strong action unless the neck is flexed against resistance. The **SCM** is innervated by the **spinal accessory nerve** (CN XI), which also innervates the **trapezius muscle.**

Other landmarks of the neck include the **laryngeal prominence** (Adam's apple) in the midline. This is formed by the **superior border of the thyroid cartilage.** The external jugular vein is prominent in some people. The external jugular vein arises from the posterior auricular and retromandibular veins just inferior to the ear and crosses over the SCM into the posterior triangle. Although variable, its usual course is to drain into the internal jugular before it joins with the subclavian vein. The **external jugular vein** is also a landmark for the **great auricular nerve**, which crosses the SCM as it ascends from the muscle's posterior border. Folds of the platysma muscle are observed when the skin over the neck is tensed (as in shaving). This muscle of facial expression is the most superficial muscle of the neck, and it courses just beneath the superficial fascia underlying the skin.

COMPREHENSION QUESTIONS

38.1 A 2-year-old girl is diagnosed with torticollis involving the right SCM. Which of the following describes the most likely anatomical change?

   A. Head flexed forward in the midline

   B. Head rotated to the right

   C. Head rotated to the left

   D. Head extended in the midline plane

38.2 A 24-year-old football player receives a blow to the left skull, and the team physician finds weakness of the left SCM. Which of the following associated findings is most likely to be seen in this patient?

   A. Weakness of the masseter muscle

   B. Decreased sensation of the ipsilateral face

   C. Decreased tearing from the ipsilateral eye

   D. Weakness of the trapezius muscle

38.3 A clinician is palpating the anterior neck of a patient who has been involved in a motor vehicle accident and notes the laryngeal prominence. Which of the following describes the anatomical structure that corresponds to this prominence?

   A. Thyroid cartilage

   B. Cricoid cartilage

   C. Hyoid bone

   D. First tracheal ring

ANSWERS

38.1 **C.** With torticollis, the SCM is shortened, leading to rotation of the head toward the contralateral side.

38.2 **D.** The SCM and the trapezius muscle are innervated by the spinal accessory nerve (CN XI), which is at risk in the posterior triangle of the neck.

38.3 **A.** The laryngeal prominence is produced by the superior border of the thyroid cartilage.

ANATOMY PEARLS

▶ The anterior neck contains structures that lie anterior to the trapezius muscle.

▶ The SCM divides the anterior neck into anterior and posterior triangles.

▶ Contraction of the SCM causes rotation of the head to the opposite side and lateral flexion.

▶ The laryngeal prominence in the midline is formed by the superior border of the thyroid cartilage

REFERENCES

Gilroy AM, MacPherson BR, Ross LM. *Atlas of Anatomy*, 2nd ed. New York, NY: Thieme Medical Publishers; 2012:512, 588–589.

Moore KL, Dalley AF, Agur AMR. *Clinically Oriented Anatomy*, 7th ed. Baltimore, MD: Lippincott Williams & Wilkins; 2014:989–992, 1007–1008.

Netter FH. *Atlas of Human Anatomy*, 6th ed. Philadelphia, PA: Saunders; 2014: plates 27, 29, 128.

CASE 39

A 67-year-old man is noted to be coughing up bright red blood for a period of 1 week. He denies exposure to tuberculosis but has smoked one pack of cigarettes per day for 30 years. On examination, his lungs are clear. Palpation of the supraclavicular regions shows a hard nontender irregular mass on the left side.

▶ What is the most likely diagnosis?
▶ What is the anatomical explanation for this particular mass?

ANSWER TO CASE 39:
Metastatic Scalene Node

Summary: A 67-year-old man who has smoked cigarettes for 30 years has a 1-week history of hemoptysis. Palpation of the supraclavicular regions shows a hard non-tender irregular mass on the left side.

- **Most likely diagnosis:** Lung cancer with a left-side supraclavicular metastatic node

- **Anatomical explanation for this particular mass:** Lymphatic drainage through the thoracic duct to the left brachiocephalic vein

CLINICAL CORRELATION

This smoker complains of **hemoptysis**, the expectoration of bright red blood, for 1 week. This is very suspicious for lung cancer. In addition, he has a hard irregular mass in the left supraclavicular region. This is most likely malignant metastasis to lymph nodes in this area. Because lymph draining the abdomen, thorax, and lower extremities is directed through the thoracic duct into the left subclavian vein, the most common location of supraclavicular node involvement is the left side.

APPROACH TO:
The Neck: Lymphatics

OBJECTIVES

1. Be able to describe general patterns of lymphatic drainage in the body
2. Be able to distinguish lymphatic flow through the supraclavicular nodes from flow through other nodes in the neck

DEFINITIONS

HEMOPTYSIS: Coughing up blood

PALPATION: Technique of physical examination that uses the hands or fingers to sense involuntary muscle tightening due to pain, or masses

METASTASIS: Spread of disease from one part of the body to another, the term generally used to describe the spread of cancer cells

DISCUSSION

The lymphatic system complements the venous system as a pathway for return of serum constituents to the heart. Blood flows from the lungs to the periphery by the pumping action of the heart. In the closed vascular system, the venous system forms from the capillaries, the vessels with the smallest diameter. The blood is drained into increasingly large veins as it is transported back to the heart and lungs. However, not all constituents of the extracellular fluid are captured into the venous system. A secondary pathway is through the lymphatic system. These fine vessels form from plexuses in tissues and, like veins, form vessels of increasing diameter. Unlike veins, however, the lymphatic vessels are not continuous channels. Instead, they are interrupted by **lymph nodes**, which contain dense aggregations of white blood cells.

In general, **lymphatics from below the diaphragm** on both sides of the body drain into the **cisterna chyli** and then to the **thoracic duct.** This is a particularly important pathway for fat droplets that are absorbed from the gastrointestinal tract after a meal. The thoracic duct ascends in the posterior mediastinum to drain into the venous system at the junction between the **left subclavian and internal jugular veins.** Above the diaphragm, including the head and neck, lymphatics on the left side also drain into the thoracic duct. On the **right side,** the vessels drain into the smaller right lymphatic duct, which drains variably into the **right subclavian vein.**

In the neck, **lymphatic vessels flow from superficial to deep, paralleling the major veins.** Several clusters of lymph nodes have been distinguished and divided into superficial and deep groups. In general, flow is from superior to inferior and from superficial to deep. However, the inferior deep group, which lies along the inferior portion of the internal jugular and subclavian veins, also drains the upper extremity, thorax, and abdomen.

The **lymphatic system is important for understanding the spread of cancer** (Figure 39-1). Unlike veins, the contractile force of the heart exerts no hydrostatic pressure in lymphatic vessels. **Lymph moves by compression from**

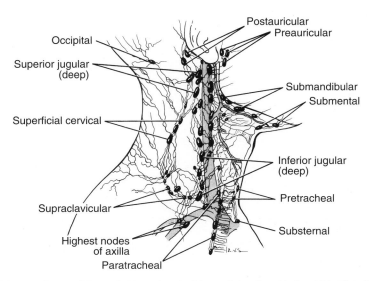

Figure 39-1. Lymphatics of the neck. (*Reproduced, with permission, from Lindner HH. Clinical Anatomy. East Norwalk, CT: Appleton & Lange, 1989:156.*)

surrounding tissues. Few lymphatic vessels have valves, so flow is highly variable. Transformed cells from one tissue can migrate through the lymphatics to adjacent tissues. Tumor cells will frequently proliferate within lymph nodes and cause the nodes to enlarge. In the neck, the supraclavicular nodes are frequently referred to as the **sentinel nodes** because their enlargement can be the first sign of cancer originating in the thorax or abdomen.

COMPREHENSION QUESTIONS

39.1 A 57-year-old man is diagnosed with colon cancer. He is noted to have a probable metastatic mass in the neck at the thoracic duct. Where is the metastasis likely to be located?

A. Right supraclavicular region

B. Right subclavicular region

C. Left supraclavicular region

D. Left subclavicular area

39.2 A 65-year-old woman is noted to have cancer of the vulva. Metastasis of the cancer to the lymph nodes in the femoral triangle is noted. Which of the following best describes the location of the lymph nodes?

A. Immediately lateral to the femoral nerve

B. Immediately medial to the femoral nerve

C. Immediately medial to the femoral artery

D. Immediately medial to the femoral vein

E. Immediately lateral to the femoral vein

39.3 Which mechanism propels lymph through the lymphatic vessels?

A. Cardiac contractility

B. Gravity

C. Peristalsis

D. Compression

ANSWERS

39.1 **C.** The thoracic duct drains into the left subclavian vein. Malignant metastasis is often diverted to the supraclavicular nodes, where it proliferates.

39.2 **D.** The relations of the structures in the groin can be recalled by the mnemonic NAVEL: Nerve, Artery, Vein, Empty space, Lymph node.

39.3 **D.** Compression is the primary means for lymph movement through the lymphatic vessels.

ANATOMY PEARLS

▶ Lymphatic vessels are not continuous channels but are interrupted by lymph nodes, which contain dense aggregations of white blood cells (lymphocytes).

▶ In general, lymphatics from below the diaphragm on both sides of the body drain into the cisterna chyli and then to the thoracic duct.

▶ The lymph ascends in the posterior mediastinum to drain into the venous system at the junction between the left brachiocephalic and internal jugular veins.

REFERENCES

Gilroy AM, MacPherson BR, Ross LM. *Atlas of Anatomy*, 2nd ed. New York, NY: Thieme Medical Publishers; 2012:128–129, 614.

Moore KL, Dalley AF, Agur AMR. *Clinically Oriented Anatomy*, 7th ed. Baltimore, MD: Lippincott Williams & Wilkins, 2014:117–118, 125, 169.

Netter FH. *Atlas of Human Anatomy*, 6th ed. Philadelphia, PA: Saunders; 2014: plates 74, 205.

CASE 40

A 28-year-old woman at 19 weeks of pregnancy complains of acute onset of numbness of the right cheek and drooping of the right face that occurred over 1 h. She denies trauma to the head. On examination, the patient has difficulty closing her right eyelid, and her right nasolabial fold is smoother than the left one. She also is drooling from the right side of her mouth. The remainder of the neurological examination is normal.

▶ What is the most likely diagnosis?
▶ What is the anatomical mechanism for this condition?

ANSWER TO CASE 40:

Bell Palsy

Summary: A 28-year-old woman at 19 weeks of pregnancy complains of acute-onset numbness of the right cheek and drooping of the right face. She denies trauma to the head. On examination, the patient has difficulty closing her right eyelid and has blunting of the right nasolabial fold. The remainder of the neurological examination is normal.

- **Most likely diagnosis:** Bell palsy (idiopathic seventh cranial nerve palsy)

- **Anatomical mechanism for this condition:** Dysfunction of the peripheral portion of the seventh cranial nerve

CLINICAL CORRELATION

Bell palsy is an idiopathic form of facial nerve paralysis that usually manifests as sudden-onset unilateral facial weakness. The peripheral portion of the facial nerve (CN VII) is affected, which may lead to loss of taste to one side of the tongue, weakness of the orbicularis oculi muscle (inability to close one's eyes), and weakness of the orbicularis oris muscle (inability to purse the lips). The upper and lower portions of the face are affected, which is consistent with a peripheral neuropathy. In contrast, lower facial weakness alone may indicate an upper motor neuron lesion. Maximal weakness usually evolves over several hours and resolves by 1 week. Although patients may experience a sensation of facial numbness, there is generally no sensory loss. Pregnancy seems to increase the incidence of Bell palsy. Keeping the eye moist and protected is an important part of therapy. The eye is vulnerable to dryness due to impaired blinking. Damage to the intracranial course of parasympathetic fibers in the greater petrosal nerve may also contribute to decreased stimulation of the lacrimal gland. Oral corticosteroid therapy may accelerate recovery. Full recovery almost always occurs.

APPROACH TO:

The Facial Nerve

OBJECTIVES

1. Be able to describe the course of the facial nerve (CN VII)

2. Be able to list the functional components of the facial nerve

DEFINITIONS

BELL PALSY: Idiopathic palsy of peripheral CN VII leading to ipsilateral facial weakness.

CHORDA TYMPANI: Small branch of the facial nerve that supplies taste receptors in the anterior two-thirds of the tongue.

UPPER MOTOR NEURON: Neurons that conduct information from motor areas of the brain to the spinal cord. The lower motor neurons project from gray matter in the spinal cord to peripheral muscle.

VIDIAN NERVE: Nerve of the pterygoid canal.

BRANCHIOMERIC MUSCLE: Skeletal muscle derived from one of the branchial arches. In general, this muscle is innervated by cranial nerves

DISCUSSION

The **facial nerve (CN VII)** originates from the **lateral surface of the caudal pons**, at the **cerebellopontine junction.** There are two roots to the nerve: the large **branchiomeric motor root** and the small **nervus intermedius**, which contains sensory and visceral motor fibers. The **facial nerve runs laterally with the vestibulocochlear nerve (CN VIII)** to enter the internal acoustic meatus (Figure 40-1). The meatus is sometimes described as having four quadrants. The facial nerve goes through the anterosuperior quadrant, whereas divisions of the vestibulocochlear nerve go through the other three.

The facial nerve continues laterally until it reaches the bony labyrinth of the inner ear. At this point, the main trunk bends sharply in a posterior direction to enter the **facial canal of the temporal bone.** The bend is called the **genu.** The fibers comprising the greater petrosal nerve arise from the genu and course anteriorly

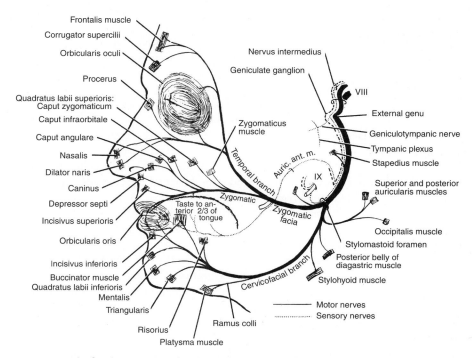

Figure 40-1. The facial nerve. (*Reproduced, with permission, from Lindner HH. Clinical Anatomy. East Norwalk, CT: Appleton & Lange, 1989:49.*)

(described further below). Also located at the genu is the **geniculate ganglion**, the cluster of sensory cell bodies that course in the facial nerve.

The facial nerve passes through the facial canal as it courses posteriorly within the **medial wall of the tympanic cavity inferior to the lateral semicircular canal.** As the canal reaches the posterior wall of the tympanic cavity, it turns inferiorly, giving off two notable branches, described below. The nerve exits the cranium through the **stylomastoid foramen**, located between the styloid and mastoid processes.

The **facial nerve** then courses anteriorly **through the parotid gland** and splits the gland into superficial and deep lobes. The nerve diverges in variable patterns to form **five major branches** that supply the muscles of facial expression: the **temporal, zygomatic, buccal, mandibular, and cervical branches.** There is also a smaller posterior auricular branch that supplies the extra-auricular muscles. Sensory nerves may innervate a small patch of skin on the posterior surface of the auricle.

The **greater petrosal nerve** emerges from the **geniculate ganglion** and courses anteriorly through a small canal. It emerges through a small hiatus into the middle cranial fossa and continues anteriorly in a groove directed toward the foramen lacerum. The nerve then passes through a tunnel in the cartilage filling the foramen or through a canal in nearby bone. After exiting the basal surface of the skull posterior to the medial pterygoid plate of the sphenoid bone, the nerve heads anteriorly through the pterygoid (Vidian) canal. The pterygoid canal courses through the sphenoid bone at the base of the medial pterygoid plate. Before entering the canal, the nerve merges with the deep petrosal nerve. The newly formed nerve of the pterygoid canal (Vidian nerve) exits anteriorly into the pterygopalatine fossa. The nerve merges with the **pterygopalatine ganglion,** which is associated with branches of the **maxillary nerve (CN V2).** Sensory and sympathetic fibers pass through the ganglion and follow the branches of the maxillary nerve throughout the nasal and oral cavities. Presynaptic parasympathetic fibers synapse in the ganglion. Postsynaptic fibers project through the same nerves to innervate glands of the oral and nasal mucosa. **Visceral motor fibers innervating the lacrimal gland** also originate in the **pterygopalatine ganglion.** These fibers run from the ganglion to the infraorbital nerve (CN V2) and follow the zygomaticotemporal nerve along the lateral wall of the orbit. They then follow the lacrimal nerve (V1) to the gland. The lacrimal nerve itself is primarily sensory and innervates the periorbital skin.

As the facial nerve descends posteriorly to the tympanic cavity, two small but important branches emerge. The first is the motor branch to the **stapedius muscle.** The belly of the stapedius is contained within the pyramid. Its tendon emerges through the apex of the pyramid to attach to the body of the stapes. Contraction of the stapedius dampens the vibration of the ossicles, thus protecting against loud sounds. The second branch in this region is the **chorda tympani.** It branches from the motor trunk before it exits the stylomastoid foramen and enters the tympanic cavity through a small canal in the posterior wall. It then runs anterolaterally, deep to the tympanic membrane. As it does so, it runs between the vertical processes of the incus and the malleus. The chorda tympani courses anteriorly and inferiorly through the temporal bone and emerges from the basal surface of the skull through the petrotympanic fissure. The nerve then courses through the infratemporal fossa along the superficial surface of the medial pterygoid muscle before joining with the lingual nerve. Sensory fibers in the chorda tympani course with branches of the lingual nerve to supply **taste receptors in the anterior two-thirds of the tongue.** Presynaptic parasympathetic fibers synapse in the **submandibular ganglion.** Postsynaptic fibers supply the **submandibular and sublingual salivary glands.**

In addition to its complex branching pattern, the facial nerve has many functional components. To summarize, the facial nerve is primarily a motor nerve that supplies branchiomeric muscles. These are primarily the **muscles of facial expression** but also include the **stapedius, stylohyoid, and posterior belly of the digastric muscle.** Another important function of the facial nerve is to supply visceral motor fibers that supply the **lacrimal gland, the submandibular and sublingual salivary glands,** and mucus-secreting **glands of the nasal and oral cavities.** The facial nerve has an important sensory component. The special sensory component that supports **taste in the anterior two-thirds of the tongue is ultimately carried by the lingual nerve.** There is a minor component of general sensation from innervation of a small patch of skin on the posterior surface of the auricle.

COMPREHENSION QUESTIONS

40.1 A 44-year-old man complains of difficulty hearing from the right ear and headaches. He also has facial muscles weakness. Which of the following is the most likely explanation?

    A. Peripheral CN VII palsy

    B. Peripheral CN VIII palsy

    C. Cerebellar pontine angle lesion

    D. Trigeminal ganglion lesion

40.2 An injury to the facial nerve (CN VII) as it leaves the stylomastoid foramen would disrupt which function?

    A. Taste to the posterior tongue

    B. Sensation to the cornea

    C. Sensation to the cheek

    D. Sensation to the anterior scalp

    E. Wrinkling of the forehead

40.3 A 33-year-old woman suffered a skull fracture that led to a unilateral facial nerve palsy. Which of the following fractures was most likely responsible?

    A. Frontal calvaria

    B. Temporal bone fracture involving the squamous part

    C. Occipital fracture

    D. Basilar fracture involving the mastoid area

ANSWERS

40.1 **C.** When multiple nerves are affected, it is unlikely to be a peripheral disorder. Cranial nerves VII and VIII exit in close proximity from the pons. A schwannoma involving the cerebellopontine angle can affect both cranial nerves.

40.2 **E.** Forehead wrinkling results from contraction of the frontalis muscle, which is innervated by the facial nerve. The facial nerve is responsible for taste in the anterior two-thirds of the tongue, but the chorda tympani emerges before the main trunk exits through the stylomastoid foramen. Sensation of the cornea and sensation to the cheek are supplied by the trigeminal nerve.

40.3 **D.** The basilar fracture involving the mastoid region of the temporal bone may impinge on the facial nerve as it exits the stylomastoid foramen.

ANATOMY PEARLS

▶ Sensory fibers in the chorda tympani course with branches of the lingual nerve to supply taste receptors in the anterior two-thirds of the tongue.

▶ The facial nerve supplies most of the muscles involved with facial expression but also supplies the stapedius, stylohyoid, and posterior belly of the digastric muscle.

▶ CN VII carries visceral motor neurons that supply the lacrimal gland, the submandibular and sublingual salivary glands, and mucus-secreting glands of the nasal and oral cavities.

REFERENCES

Gilroy AM, MacPherson BR, Ross LM. *Atlas of Anatomy*, 2nd ed. New York, NY: Thieme Medical Publishers; 2012:488–489, 492–493, 504–505, 514–515, 528–530.

Moore KL, Dalley AF, Agur AMR. *Clinically Oriented Anatomy*, 7th ed. Baltimore, MD: Lippincott Williams & Wilkins; 2014:853–855, 861, 1068–1070.

Netter FH. *Atlas of Human Anatomy*, 6th ed. Philadelphia, PA: Saunders; 2014: plates 24, 124.

A 35-year-old woman complains of excruciatingly painful spasms in the right cheek and chin. These pain episodes last for a few seconds and are intense. She was diagnosed with multiple sclerosis 2 years previously. She is not taking medications currently, although she previously received intravenous corticosteroid therapy. Her physician says that her problem is related to the nerve that innervates the skin of the cheek area.

▶ What is the most likely diagnosis?
▶ What is the anatomical explanation for this condition?

ANSWER TO CASE 41:

Trigeminal Neuralgia

Summary: A 35-year-old woman who has multiple sclerosis complains of excruciating spasms of pain that affect her right cheek and chin and last for a few seconds.

- **Most likely diagnosis:** Trigeminal neuralgia.

- **Anatomical explanation for this condition:** Pain follows the distribution of CN V, which innervates the eyes, cheeks, and chin.

CLINICAL CORRELATION

Trigeminal neuralgia (tic douloureux) is among the most excruciating types of pain seen by clinicians and is so intense that it will cause the patient to wince. This young woman complains of several seconds of intense spasmodic pain of the right cheek and chin. Her history of multiple sclerosis is important because trigeminal neuralgia is relatively common in this group of patients. The character of the pain excludes some of the other common etiologies of head or facial pain such as migraine headache (usually throbbing unilateral pain with orbital involvement) or tension headache (bandlike constricting pain from the temples to the occiput bilaterally). She has no history of herpes simplex virus, which can also affect CN V. CN V has three branches of sensory distribution. Treatment includes carbamazepine or baclofen and, in severe cases, trigeminal nerve ablation.

APPROACH TO:

Trigeminal Nerve

OBJECTIVES

1. Be able to relate the dermatomes of the face to the branches of the trigeminal nerve (CN V)

2. Be able to list the functions of the trigeminal nerve

DEFINITIONS

MULTIPLE SCLEROSIS: Disease in which plaques in the nervous system arise because of the proliferation of fibrous connective tissue or glial cells. **Sclerosis** in general refers to hardening of the tissue, as in atherosclerosis, or hardening of the arteries.

BACLOFEN: Muscle relaxing drug that acts through type b γ-aminobutyric acid (GABA$_b$) receptors.

CARBAMAZEPINE: Centrally acting anticonvulsive drug of unknown action.

DISCUSSION

The **trigeminal nerve** exits the brain from the **lateral surface of the pons.** The sensory fibers arise as a large root. Motor fibers to the **muscles of mastication** usually arise as a separate smaller root. The nerve courses on the lateral surface of the sphenoid bone deep to the cavernous sinus. The cell bodies of the sensory nerves form the **trigeminal ganglion** along the medial wall of the middle cranial fossa. Three large nerves emerge from the ganglion: the **ophthalmic, maxillary, and mandibular divisions** of the trigeminal nerve (Figure 41-1).

Branches of these nerves supply general sensation to the face and anterior scalp. The posterior scalp is supplied by cervical spinal nerves. The **ophthalmic nerve** supplies the dermatome that courses superiorly to the horizontal midline of the orbit. It also supplies the midline region of the nose. The **maxillary nerve** supplies the region over the maxilla, inferior to the orbit, including the lateral surface of the nose and the upper lip. A small band extends superiorly over the zygomatic arch and temporalis muscle. The mandibular division supplies a band of skin running superiorly over the temporalis muscle. The major branches of the ophthalmic nerve that supply skin are the supraorbital and supratrochlear nerves, which supply skin of the forehead and anterior scalp. The nasociliary nerve supplies skin over the medial nose through the external nasal branch of the anterior ethmoidal nerve.

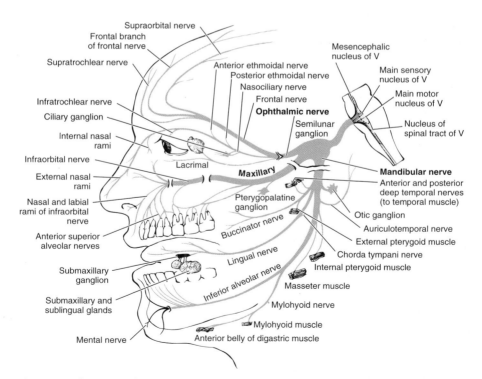

Figure 41-1. The trigeminal nerve. (*Reproduced, with permission, from Waxman SG. Clinical Neuroanatomy, 25th ed. New York: McGraw-Hill, 2003:112.*)

The maxillary nerve innervates the skin primarily through the infraorbital nerve. More laterally, the **zygomaticofacial and zygomaticotemporal nerves** also contribute. The branches of the mandibular nerve that innervate the skin are the auriculotemporal superiorly and the mental nerve (a branch of the inferior alveolar) inferiorly. The buccal nerve supplies skin over the cheek. This nerve also innervates buccal mucosa in the oral cavity. Although its branches pass through the buccinator, they do not provide motor innervation. The **buccinator** is supplied by the buccal branch of the **facial nerve (CN VII).**

COMPREHENSION QUESTIONS

41.1 A 56-year-old man had a stroke. Among other symptoms, a marked deficit in bite strength was observed on the affected side, indicating weakness in the muscles of mastication. Which of the following muscles is also innervated by the same nerve?

    A. Orbicularis oculi

    B. Platysma

    C. Anterior belly of digastric

    D. Stylohyoid

    E. Superior belly of omohyoid

41.2 A 45-year-old diabetic woman has developed shingles involving the right cornea. Through which nerve did the varicella virus most likely travel to the cornea?

    A. CN II

    B. CN III

    C. CN V

    D. CN VII

41.3–41.6 Match the following divisions (A–C) to the branches 41.3–41.6.

    A. CN V1

    B. CN V2

    C. CN V3

41.3 Auriculotemporal nerve

41.4 Lacrimal nerve

41.5 Supraorbital nerve

41.6 Infraorbital nerve

ANSWERS

41.1 **C.** The anterior belly of digastric is innervated by CN V3, as are the muscles of mastication. The platysma and orbicularis oculi muscles are supplied by CN VII.

41.2 **C.** The trigeminal nerve supplies sensory innervation to the cornea. Herpes simplex infections or varicella virus infections involving the face may travel through CN V to the cornea and endanger vision.

41.3 **C.** CN V3 supplies the auriculotemporal, buccal, and mental nerves.

41.4 **A.** CN V1 supplies the lacrimal, supraorbital, and supra- and infratrochlear nerves.

41.5 **A.** CN V1 supplies the lacrimal, supraorbital, and supra- and infratrochlear nerves.

41.6 **B.** CN V2 supplies the infraorbital and zygomaticotemporal nerves.

ANATOMY PEARLS

► The trigeminal nerve (CN V) exits the brain from the lateral surface of the pons.

► The trigeminal nerve comprises three divisions: ophthalmic, maxillary, and mandibular.

► Branches of these nerves supply general sensation to the face and anterior scalp. The posterior scalp is supplied by cervical spinal nerves.

REFERENCES

Gilroy AM, MacPherson BR, Ross LM. *Atlas of Anatomy*, 2nd ed. New York, NY: Thieme Medical Publishers; 2012:494–495, 502–503, 514–515, 530–533.

Moore KL, Dalley AF, Agur AMR. *Clinically Oriented Anatomy*, 7th ed. Baltimore, MD: Lippincott Williams & Wilkins; 2014:849–853, 1065–1067, 1081.

Netter FH. *Atlas of Human Anatomy*, 6th ed. Philadelphia, PA: Saunders; 2014: plates 2, 12, 52, 122–123.

A 38-year-old male presents to the emergency room with complaints of a persistent headache and problems with his left eye. He has no known medical problems, and his headache is slightly improved with ibuprofen. He denies having any previous vision problems. Examination reveals ptosis, dilated pupil, and displacement "down and out" in his left eye The remainder of the exam is normal. An MRI shows an aneurysm of the circle of Willis.

- ▶ Given the physical exam, what ocular muscles are likely to have been unaffected?
- ▶ Which nerve is likely to have been affected?

ANSWER TO CASE 42:
Oculomotor Nerve Palsy

Summary: A 38-year-old healthy male with recent-onset left eye findings of ptosis, dilated pupil, and displacement of eye "down and out."

- **Ocular muscles not involved:** Superior oblique and lateral rectus

- **Nerve affected:** Oculomotor nerve (CN III)

CLINICAL CORRELATION

Findings of ptosis, dilated pupil, and down-and-out eye displacement are most consistent with oculomotor nerve palsy. The oculomotor nerve is the third of 12 paired cranial nerves and originates from the midbrain. It controls most eye movements, constriction of the pupil, and eyelid position. Down-and-out displacement of the eye occurs from the unopposed action of the lateral rectus and superior oblique. The superior oblique muscle is innervated by the trochlear nerve (CN IV), and the lateral rectus muscle is innervated by the abducens nerve (CN VI). An oculomotor nerve palsy may be caused by an aneurysm, compression, infection, infarction, or tumor.

APPROACH TO:
Extraocular Muscles of the Orbit

OBJECTIVES

1. Be able to name the seven extraocular eye muscles of each orbit, and also their attachments, actions, and innervation

2. Be able to describe how each of these muscles is optimally tested in a clinical setting

3. Be able to describe the presentation of a patient with injury to each nerve that innervates these muscles

DEFINITIONS

PTOSIS: Drooping or partial closure of the upper eyelid

NERVE PALSY: Partial or incomplete paralysis

ANEURYSM: Dilatation of the wall of an artery due to an acquired or congenital condition

Table 42-1 • EXTRAOCULAR MUSCLES OF THE ORBIT				
Muscle	Origin	Insertion	Action	Innervation
Superior oblique	Posterior orbit	Posterosuperior sclera	Depresses and abducts the eye	CN IV
Inferior oblique	Anterior orbital floor	Posteroinferior sclera	Elevates and abducts the eye	CN III
Superior rectus	The common tendinous ring	The anterior portion of sclera	Elevates and adducts	CN III
Inferior rectus	The common tendinous ring	The anterior portion of sclera	Depresses and adducts	CN III
Lateral rectus	The common tendinous ring	The anterior portion of sclera	Abducts	CN VI
Medial rectus	The common tendinous ring	The anterior portion of sclera	Adducts	CN III

DISCUSSION

The extraocular muscles of the orbit are the levator palpebrae superioris, four rectus (superior, inferior, lateral, and medial), and two oblique (superior and inferior) muscles. All of the extraocular muscles originate from the apex of the pyramidal shaped orbit near the optic canal, except for the inferior oblique muscle, which arises from the anterior orbital floor. The levator superioris attaches directly to the eyelid and controls its movements. Rarely do any of the six muscles attaching directly to the eyeball move the eyeball independently from the other muscles, although their individual actions are typically described. Their attachments, actions, and innervation are listed in Table 42-1 and illustrated in Figure 42-1.

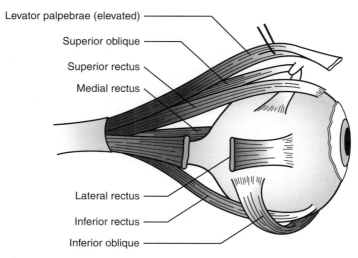

Levator palpebrae (elevated)
Superior oblique
Superior rectus
Medial rectus
Lateral rectus
Inferior rectus
Inferior oblique

Figure 42-1. Diagram of eye muscles.

The **levator palpebrae superioris** muscle originates from the posterior orbit and attaches to the skin and tarsal plate of the upper eyelid, which it elevates. It is opposed by the superior portion of the orbicularis oculi muscle. It contains smooth muscle fibers forming the **superior tarsal** muscle that is innervated by sympathetic nerve fibers during fright or startle responses.

The **superior oblique** muscle originates anatomically from the posterior apex of the orbit and passes anteriorly to the **trochlea,** a pulley like fibrous ring at the super-omedial margin on the orbit. The trochlea is the functional origin of this muscle. Its tendon passes through the trochlea to insert on posterosuperior portion of the sclera. On contraction, it pulls the posterior portion of the eyeball anteriorly and medially. Thus **the pupil is turned down and out.** The **inferior oblique** muscle originates from the anteromedial floor of the orbit, thus simulating the portion of the superior oblique between the trochlea and insertion. It inserts into the posteroinferior sclera and therefore opposes the action of the superior oblique. It will turn the pupil of the eye up and out. The two oblique muscles also produce extorsion or lateral rotation of the eyeball.

The four rectus muscles (superior, inferior, lateral, and medial) all originate from a common tendinous ring surrounding the optic canal and a portion of the superior orbital fissure in the posterior orbit. Each inserts on the anterior half of the sclera on that portion of the eyeball according to their name. Thus the lateral rectus inserts on the anterolateral sclera. Note that the superior and inferior rectus muscles will turn the eyeball in or adduct the pupil and will also produce intorsion or medial rotation of the eyeball.

For the sake of clarity, the following descriptions for muscle testing are for **only the right eye.** For optimal testing of the extraocular muscles, the axis of the muscle is placed parallel with the axis of muscle pull. With the eyeball (pupil) abducted, the superior and inferior recti are in line with their pull, and their action on the eyeball is almost purely elevation and depression, respectively. For the superior and inferior oblique muscles, adduction of the eyeball (pupil) places the axis of the muscle in line with its pull (remember that the functional origin of the superior oblique is the trochlea). Thus the eyeball (pupil) is depressed and elevated for these two muscles. The lateral and medial recti are tested by simply adducting or abducting the eyeball (pupil) (see Figure 42-2).

If the **oculomotor nerve** (CN III) is injured as in this case, the pupil of the affected eye will be turned down and out (because of the unopposed action of the superior oblique and lateral rectus muscle). The pupil will also be dilated because of loss of the parasympathetic innervation to the constrictor muscle of the pupil. Loss of the **trochlear nerve,** although rare, results in slight adduction of the affected eye, weakness of downward gaze due to paralysis of the superior oblique, and head tilting to eliminate diplopia (double vision). Loss of **abducens nerve** (CN VI) function results in paralysis of the lateral rectus muscle, and thus the patient's affected eye will be turned in or adducted.

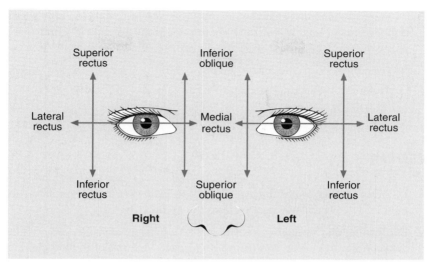

Figure 42-2. Diagram of eye muscle action.

COMPREHENSION QUESTIONS

42.1 While examining a patient, you notice ptosis of the patient's left eye. This would indicate to you there is paralysis of which muscle?

A. Orbicularis muscle

B. Superior oblique muscle

C. Inferior oblique muscle

D. Inferior rectus muscle

E. Levator palpebrae muscle

42.2 While performing a physical examination, you test the function of the muscles attached to the eyeball and thereby the integrity of their innervation. You would test the function of the superior oblique muscle by having the patient do which of the following?

A. Look in toward the nose

B. Look laterally

C. Look in toward the nose and then upward

D. Look in toward the nose and then down

E. Look laterally and then down

42.3 During this same physical exam, you ask the patient to look laterally with her right eye, and then upward. You have just tested the function of which muscle?

A. Superior rectus muscle

B. Superior oblique muscle

C. Inferior oblique muscle

D. Inferior rectus muscle

E. Medial rectus muscle

ANSWERS

42.1 **E.** Ptosis or drooping of the eyelid is due to paralysis of the levator palpebrae muscle. The orbicularis muscle closes the eyelid.

42.2 **D.** Turning the eyeball inward places the portion of the superior oblique between the trochleae, and its insertion places the axis of the muscle in line with its axis of muscle pull. Because the muscle's insertion is on the posterior portion of the sclera, it will then turn the eye (pupil) down (depress).

42.3 **A.** Turning the eyeball out places the axis of the superior rectus muscle parallel to its pull, and the muscle will then turn the eyeball upward (elevate).

ANATOMY PEARLS

▶ $LR_6SO_4AO_3$: lateral rectus, CN **VI**; superior oblique, CN **IV**; all others, CN **III**.

▶ The functional portion of the superior oblique muscle is between the trochlea and the insertion of the tendon.

▶ The oculomotor nerve (CN III) innervates the majority of the extraocular muscles, the sphincter muscle of the pupil, and the smooth muscle fibers of the superior tarsal muscle.

REFERENCES

Gilroy AM, MacPherson BR, Ross LM. *Atlas of Anatomy*, 2nd ed. New York, NY: Thieme Medical Publishers; 2012:501, 538–539, 541–543.

Moore KL, Dalley AF, Agur AMR. *Clinically Oriented Anatomy*, 7th ed. Baltimore, MD: Lippincott Williams & Wilkins; 2014:898–899, 903–905, 913.

Netter FH. *Atlas of Human Anatomy*, 6th ed. Philadelphia, PA: Saunders; 2014: plates 86, 88, 122.

CASE 43

A male infant who weighs 3500 g appears icteric on examination. The previous day, the infant was delivered vaginally by vacuum-assisted extraction because there were severe fetal heart rate decelerations. The infant's scalp has a 5-cm discolored soft tissue swelling that seems to be contained by and does not cross the sagittal or lambdoidal sutures. The mother had no prenatal or medical problems. There is no family history of bleeding disorders.

▶ What is the most likely diagnosis?
▶ What is the anatomical mechanism for the condition?

ANSWER TO CASE 43:

Cephalohematoma

Summary: On the previous day, a 3500-g male infant was delivered vaginally by vacuum-assisted extraction. The infant appears icteric, and his scalp has a 5-cm hematoma that is contained by and does not cross the sagittal or lambdoidal sutures.

- **Most likely diagnosis:** Cephalohematoma

- **Anatomical mechanism for the condition:** Injury to the branches of arteries supplying the lateral skull

CLINICAL CORRELATION

This 1-day-old infant was delivered with the aid of vacuum extraction and now has icterus and a discolored soft tissue mass that is contained within the sutures. This almost certainly represents a cephalohematoma. The more common **caput succedaneum**, which is swelling of the scalp soft tissue, is a normal response of the fetal head to the birth process. In this situation, the blood will cross over suture lines. When a soft tissue mass seems contained by suture lines, **subgaleal** cephalohematoma is suspected. The hemoglobin deposited in the hematoma becomes bilirubin, which is the reason for the infant's icterus. A skull radiograph or CT scan is usually obtained to assess for skull fracture. Most of these hematomas will resolve with observation.

APPROACH TO:

The Scalp and Skull

OBJECTIVES

1. Be able to define the layers of the scalp

2. Be able to describe the structure of cranial sutures

DEFINITIONS

MAJOR SUTURES OF THE SKULL: The **sagittal suture** runs along the midline of the skull between the two parietal bones. The **lambdoidal suture** runs left to right posteriorly and separates the two parietal bones from the occipital bone. The **coronal suture** has the same course anteriorly and separates the frontal bone from the two parietal bones.

HEMATOMA: Pool of blood that accumulates in a tissue or space, usually clotted.

BILIRUBIN: Bile salt that is formed from the breakdown of hemoglobin by the liver. It usually accumulates in the gallbladder and is excreted into the small bowel to facilitate digestion. High levels of bilirubin in the blood give the skin and sclera of the eyeballs a yellowish tint (jaundice).

ICTERIC: Yellowish appearance of jaundice, in this case due to the local breakdown of bilirubin in the blood that has accumulated in the hematoma.

CAPUT SUCCEDANEUM: Edematous swelling of the superficial scalp due to the normal trauma of the birth process that resolves within 2 to 3 days.

DISCUSSION

The scalp is the unit of tissue that covers the calvaria. The tissue is composed of **five layers** and can be remembered by the acronym **SCALP** (Figure 43-1). Most superficial is the **skin**, which includes the dermis and the superficial fascia. Deep to that is a layer of dense **connective tissue** that binds tightly to the skin. The next layer is the **aponeurosis** of the occipitofrontalis muscle (galea aponeurotica). These three layers adhere tightly and move together as a unit. The fourth layer consists of **loose connective tissue.** The fifth layer is the **periosteum**, which covers the bone itself. The periosteum adheres tightly to the bone, especially in the region of

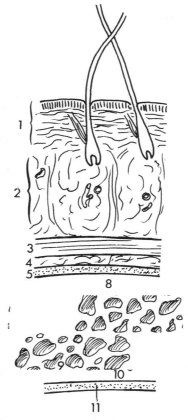

Figure 43-1. The layers of the scalp: 1 = skin, 2 = connective tissue, 3 = aponeurosis, 4 = loose connective tissue, 5 = periosteum, 8 = outer table of calvaria, 9 = diploë, 10 = inner table of calvaria, 11 = endocranium. (*Reproduced, with permission, from the University of Texas Health Science Center, Houston Medical School.*)

the cranial sutures. The flexibility of the loose connective tissue allows the more superficial layers to move over the periosteum. In infants, the periosteum is adherent to the sutures.

The blood vessels that supply the scalp arise from branches of the **internal** and **external carotid arteries. Anteriorly**, these are the **supraorbital and supratrochlear arteries, which are derived from the internal carotid.** More **laterally and posteriorly**, the scalp is supplied by branches of the **external carotid arteries.** These include the **superficial temporal arteries**, which ascend in front of the auricle; and the **occipital** and **posterior auricular arteries,** which ascend posterior to the auricle. **The arteries of the scalp are highly anastomotic.** Therefore, after a laceration, blood may pulse from both ends of the cut artery.

The **nerves of the scalp anteriorly** originate from the **first and third divisions of the trigeminal nerve. Medially**, the **supraorbital and supratrochlear nerves** supply sensory innervation. Laterally, the **auriculotemporal nerve** provides sensory innervation. The posterior scalp is supplied medially by the posterior primary rami of cervical spinal nerves (C2, as the greater occipital nerve, and C3). Laterally, the skin is supplied by anterior primary rami that form the cervical plexus, in particular the **lesser occipital and posterior auricular nerves.**

Trauma to the scalp can damage blood vessels and hence cause a **hematoma.** The hematoma may spread within the same layer. Blood in the superficial fascia will migrate a little more slowly because of the septa within the subcutaneous fascia. In newborn infants, hematomas in this layer commonly result from the trauma of movement through the birth canal. Likewise, scalp trauma such as that induced by a suction-assisted delivery may occasionally injure the **arteries within the periosteum, leading to accumulation of blood between the periosteum and the bone.** Because the periosteum in infants adheres to the sutures, spread is impeded. A subcutaneous hematoma will cross sutures, but a subperiosteal hematoma will not. In adults, the **loose connective layer** is called the "danger space" because infection can easily migrate to the **periorbital space.** This is a dangerous condition because of the potential for spread into the cranium through the **cavernous sinus.**

COMPREHENSION QUESTIONS

43.1 Which of the following best describes the layers of scalp?

 A. Skin, aponeurosis, dense connective tissue, periosteum

 B. Skin, loose connective tissue, aponeurosis, periosteum

 C. Skin, dense connective tissue, aponeurosis, loose connective tissue, periosteum

 D. Skin, aponeurosis, loose connective tissue, muscle, periosteum

43.2 A 65-year-old woman complains of severe pain of the right side of the head. A vascular surgeon takes a biopsy of the artery deep to the temporalis muscle. Which of the following vessels did the surgeon most likely biopsy?

   A. Middle meningeal artery
   B. External carotid artery
   C. Ophthalmic artery
   D. Deep temporal artery

43.3 A neurologist uses a pin to test the sensation to a 26-year-old man's scalp just near the hair line anteriorly. Which of the following nerves provides the innervation to the scalp in this region?

   A. CN V
   B. CN VII
   C. CN X
   D. Spinal nerve C2

ANSWERS

43.1 **C.** The layers of the scalp can be remembered by the mnemonic SCALP: Skin, Connective tissue, Aponeurosis, Loose connective tissue, Periosteum.

43.2 **D.** The temporal artery is deep to the temporalis muscle and sometimes is associated with inflammation (temporal arteritis). Temporal arteritis or giant cell arteritis is associated with headache and multiple joint pain.

43.3 **A.** The anterior scalp is innervated by CN V, whereas the posterior scalp is innervated by spinal nerve C2.

ANATOMY PEARLS

▶ The blood vessels that supply the scalp are from branches of the internal and external carotid arteries.

▶ The sensory innervation of the scalp is by the trigeminal nerve: anteriorly by the supraorbital and supratrochlear nerves and laterally by the auriculotemporal nerve. The posterior scalp is innervated by spinal nerves C2 and C3. Spinal nerve C1 does not have a sensory component.

▶ The arteries of infants who undergo trauma, such as that induced by a suction-assisted delivery, may be damaged within the periosteum and develop a subperiosteal hematoma.

REFERENCES

Gilroy AM, MacPherson BR, Ross LM. *Atlas of Anatomy,* 2nd ed. New York, NY: Thieme Medical Publishers; 2012:488–489, 516–517, 528–529.

Moore KL, Dalley AF, Agur AMR. *Clinically Oriented Anatomy,* 7th ed. Baltimore, MD: Lippincott Williams & Wilkins; 2014:843–844, 856, 860–861.

Netter FH. *Atlas of Human Anatomy,* 6th ed. Philadelphia, PA: Saunders; 2014: plates 3, 14.

A 15-year-old boy was the pitcher for his little league baseball team when he was hit by a line drive to the right temple area. He lost consciousness briefly but woke up after about 45 s and had no neurological deficits. He was taken to the emergency room and seemed to be in good condition. Four hours later, while being observed, he complained of an increasing headache and had a seizure. On examination, the patient's right pupil appeared dilated and reacted sluggishly to light. The emergency physician is concerned about increased intracranial pressure.

▶ What is the most likely diagnosis?
▶ What is the anatomical explanation for this condition?

ANSWER TO CASE 44:
Epidural Hematoma

Summary: A 15-year-old boy was hit by a baseball to the right temple area. He lost consciousness briefly and had a lucid interval. Four hours later, he developed an increasing headache, a dilated and sluggish right pupil, and had a seizure, consistent with increased intracranial pressure.

- **Most likely diagnosis:** Epidural hematoma resulting in increased intracranial pressure

- **Anatomical explanation for this condition:** Disruption of a branch of the middle meningeal artery, which causes a growing hematoma between the dura and cranium and puts pressure on the underlying brain

CLINICAL CORRELATION

This 15-year-old baseball player underwent significant blunt trauma to the right temple area by a baseball. He briefly lost consciousness, likely due to the concussion of the baseball. After waking up, he had no neurological deficits; however, after 4 h, there were signs of increased intracranial pressure. The most likely explanation is disruption of the middle meningeal artery, which underlies the temporal bone. Over time, the hematoma formed, putting pressure on the underlying brain tissue. The ipsilateral pupil was affected by compression of the oculomotor nerve (CN III) by the temporal lobe of the brain. This scenario of a loss of consciousness followed by a lucid interval and a second loss of consciousness is very typical for epidural hematoma. Because this is arterial bleeding, rapid expansion of the hematoma is typical. Emergent cerebral decompression and surgical control of the bleeding are paramount.

APPROACH TO:
Meninges and Arterial Supply to Brain

OBJECTIVES

1. Be able to list the meningeal layers

2. Be able to identify the dural folds and associated dural sinuses

3. Be able to describe the vascular supplies to the meninges and underlying brain

DEFINITIONS

PACHYMENINX: The thick meningeal layer, that is, the dura mater.

LEPTOMENINX: The thin meningeal layers, that is, the arachnoid and pia maters together.

DURAL SINUS: Cavity filled with venous blood formed by a split in the two layers of dura mater, the periosteal and meningeal layers. Blood drains from the system of sinuses into the internal jugular vein.

PTERION: A landmark on the lateral surface of the skull formed by the junction of the frontal, parietal, temporal, and sphenoid bones. It usually has an H-shaped appearance.

DISCUSSION

As in the spinal cord, three meningeal layers cover the brain: the dura mater, arachnoid mater, and pia mater. The **dura mater** is a thick, strong membrane (pachymeninx) that is closely apposed to the deep surface of the cranium. Immediately deep to the dura is the arachnoid layer, a thin, nearly transparent membrane that adheres to the deep surface of the dura. The arachnoid layer is separated from the brain by the **subarachnoid space,** which is filled with CSF. The **pia mater** is a thin layer attached to the surface of the brain itself. The **arachnoid and pia layers together** may be referred to as the **leptomeninges.**

The **dura mater** that covers the external surface of the brain consists of **two layers, an external periosteal layer attached to the bone and an internal meningeal layer.** The internal layer forms folds that separate the major lobes of the brain. The **falx cerebri** courses along the midline and separates the left and right cerebral hemispheres. Running at right angles, the **tentorium cerebelli** separates the two lobes of the cerebrum from the cerebellum. On the inferior surface of the tentorium is attached the small **falx cerebelli**, which also runs along the midline and partially separates the cerebellum into lobes. Another important dural infolding the **diaphragma sellae,** covers the pituitary fossa.

Normally, the two dural layers are tightly apposed, but they may split to form the **dural sinuses** (Figure 44-1). The major dural sinuses are **the superior sagittal sinus,** which courses along the superior edge of the falx cerebri, and the **transverse sinus,** which courses along the posterior border of the tentorium cerebelli. The **transverse sinus** continues laterally as the **sigmoid sinus** and empties into the internal jugular vein. On the inferior surface of the falx cerebri, the **inferior sagittal sinus** continues as the straight sinus after joining the **great vein of Galen,** which drains the brain. The **superior, straight, and transverse sinuses come together at the confluence of sinuses,** a landmark on the internal surface of the occipital bone. Other important sinuses are the superior and inferior petrosal sinuses and the cavernous sinus.

The vessels that supply the dura mater are branches of the **middle meningeal artery.** This artery arises in the infratemporal fossa from the first part of the maxillary artery and enters the cranial cavity through the foramen spinosum. The artery runs within the dura mater and separates into anterior and posterior divisions. An external landmark for the middle meningeal artery is the **pterion, where the frontal, parietal, temporal, and sphenoid bones converge.** The vessels that supply the brain arise from the **circle of Willis** (see Case 46 for more details). This **anastomotic formation originates from the internal carotid and vertebral arteries.** The major branches tend to course along the surface of the brain and give off penetrating branches.

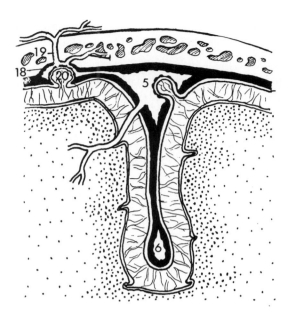

Figure 44-1. The dura and meninges: 5 = superior sagittal sinus, 6 = inferior sagittal sinus, 18 = lateral lacuna, 19 = emissary vein (connects sinuses to scalp veins), 20 = arachnoid granulation (resorption of cerebrospinal fluid). (*Reproduced, with permission, from the University of Texas Health Science Center, Houston Medical School.*)

Head trauma can result in damage to vessels and internal bleeding. Blood accumulates in potential spaces surrounding the brain, expanding their volume, and placing pressure on the brain. The site of accumulation is characteristic of the type of vessel that is damaged. For example, rupture of the **middle meningeal artery** will lead to accumulation of blood in the **epidural potential space, between the external periosteal layer of the dura and the calvaria.** Blood from a **cerebral artery** due, for example, to a **ruptured cerebral aneurysm**, will accumulate in the **subarachnoid space.** Head trauma may result in **rupture of veins as they enter a sinus**, usually resulting in accumulation of blood in **the subdural potential space between the dural and arachnoid layers.** These veins may be cerebral veins that drain the brain or emissary veins that drain the scalp.

COMPREHENSION QUESTIONS

44.1 A 35-year-old man developed an intracranial hemorrhage when one of the meningeal arteries ruptured. Anatomically, where is the hematoma located?

A. Immediately superficial to the dura

B. Immediately deep to the dura

C. Within the subarachnoid space

D. Within the brain parenchyma

44.2 A 1-month-old infant is seen in the emergency department because of lethargy and seizures. After careful questioning, it was discovered that the infant was shaken before the change in mental status. Which vessels are most likely to be injured?

   A. Meningeal arteries

   B. Meningeal veins

   C. Emissary veins

   D. Middle cerebral arteries

44.3 A 21-year-old man is brought into the emergency room after being hit in the head with a baseball bat. The neurosurgeon suspects that the skull fracture and underlying hematoma occurred at the junction of the four major bones of the skull. Which of the following describes this region?

   A. Bregma

   B. Lambda

   C. Pterion

   D. Nasion

ANSWERS

44.1 **A.** Injuries to the meningeal arteries lead to epidural hematomas.

44.2 **C.** Infants who are shaken are vulnerable to laceration of the emissary veins that are found below the dura. Thus, they often develop subdural hematomas.

44.3 **C.** The pterion is a landmark of the skull where the four major bones of the skull (frontal, parietal, temporal, and sphenoidal) come together. It is also the thinnest part of the skull.

ANATOMY PEARLS

► The dura mater, which covers the external surface of the brain, consists of two layers, an external periosteal layer attached to the bone and an internal meningeal layer.

► An external landmark for the middle meningeal artery is the pterion, where the frontal, parietal, temporal, and sphenoid bones converge.

► The vessels that supply the dura mater are branches of the middle meningeal artery; injuries to these vessels lead to epidural hematomas.

► Blood from ruptured cerebral arteries, such as those due to a ruptured cerebral aneurysm, will accumulate in the subarachnoid space.

REFERENCES

Gilroy AM, MacPherson BR, Ross LM. *Atlas of Anatomy*, 2nd ed. New York, NY: Thieme Medical Publishers; 2012:524–525, 634–635.

Moore KL, Dalley AF, Agur AMR. *Clinically Oriented Anatomy*, 7th ed. Baltimore, MD: Lippincott Williams & Wilkins; 2014:865–874, 876–877.

Netter FH. *Atlas of Human Anatomy*, 6th ed. Philadelphia, PA: Saunders; 2014: plates 101–103.

A 36-year-old woman complains of pain and swelling beneath the left mandible, particularly after eating a meal. On examination, she is noted to have edema and tenderness of the left submandibular region. Palpation of her mouth reveals a 4-mm, irregular, nonmobile, hard mass in the mucosa of her mouth. She denies trauma to the region and does not have an eating disorder.

▶ What is the most likely diagnosis?
▶ What is the anatomical course of the affected structure?

ANSWER TO CASE 45:

Salivary Stone

Summary: A 36-year-old woman complains of pain and swelling of the left subman-dibular area. On examination, she has tenderness of the left submandibular salivary gland and a palpable, irregular, 4-mm mass along the floor of her mouth. She denies trauma to the region and does not have an eating disorder.

- **Most likely diagnosis:** Stone in the submandibular duct (sialolithiasis).

- **Anatomical course of the affected structure:** The submandibular salivary duct drains from the deep lobe of the submandibular gland and courses anterolaterally along the base of the tongue. Occlusion of the duct by a stone will cause secreted saliva to accumulate proximally to the stone, thus causing distention and pain.

CLINICAL CORRELATION

This 36-year-old woman has sudden onset of pain to the left submandibular area. The pain is most intense after a meal. She also complains of a "sandlike" or "gritty" sensation in her mouth. The left submandibular gland appears swollen. This is most consistent with a stone in the submandibular duct. Pain after a meal is due to the accumulation of saliva proximal to the occluded duct, which stretches the duct or the capsule of the gland. Generalized swelling may be due to a secondary infection. The pathogenesis of sialolithiasis is unknown but appears to be due to lodging of a small particle in the duct, which serves as a nucleus for deposition of organic and inorganic material. The particle could be food, bacteria, or an inorganic constituent of tobacco smoke. The next diagnostic step would be examination with sialoendoscopy. Treatment would be excision of the stone under endoscopy and administration of antibiotics. If necessary, the gland would be removed surgically.

APPROACH TO:

Salivary Glands

OBJECTIVES

1. Be able to describe the salivary glands and the course of their ducts to the oral cavity

2. Be able to identify structures in the floor of the mouth related to the subman-dibular gland

DEFINITIONS

CARUNCLE: Small protuberance, or papilla

FRENULUM: Mucosal fold that extends along the midline from the floor of the mouth to the inferior surface of the tongue

DISCUSSION

Three salivary glands form an irregular, space-filling ring around the oral cavity (Figure 45-1). The **parotid gland** lies superficial and posterior to the ramus of the mandible and inferior to the ear. The **submandibular gland** lies below the angle and the body of the mandible superficial to the mylohyoid muscle. The **sublingual gland** lies in the floor of the mouth between the mandible and the genioglossus muscle. All of the glands secrete saliva into the oral cavity through characteristic ducts. The parotid duct emerges from the anterior border of the parotid gland. The **parotid duct crosses over the masseter muscle** and pierces through the buccinator muscle to open into the oral cavity, typically at the level of the second upper molar. The **submandibular duct** forms from the deep lobe of the submandibular gland, **deep to the mylohyoid muscle.** The duct runs anteriorly on the surface of the hyoglossus muscle and opens into the oral cavity through the submandibular caruncles, just lateral to the lingual frenulum. The **sublingual glands** give rise to numerous small ducts that **empty at the base of the tongue.**

The **submandibular duct** has a close relation to several important structures in the floor of the mouth. The submandibular gland folds around the free posterior border of the **mylohyoid muscle**, and the duct arises from the deep lobe of the gland. It courses anteriorly between the mylohyoid and hyoglossus muscles and then on

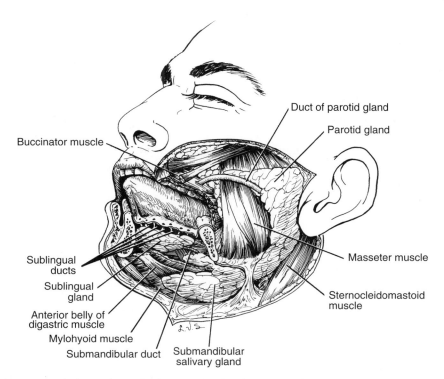

Figure 45-1. The large salivary glands. (*Reproduced, with permission, from Lindner HH. Clinical Anatomy. East Norwalk, CT: Appleton & Lange, 1989:60.*)

the deep surface of the sublingual gland. The **hypoglossal nerve (CN XII) courses inferior to the submandibular duct** to enter the inferior surface of the genioglossus muscle. The lingual nerve descends on the surface of the medial pterygoid muscle and loops underneath the duct before innervating the anterior portion of the tongue.

COMPREHENSION QUESTIONS

45.1 A 22-year-old male is involved in a knife fight after a soccer game. He is brought to the emergency department. An 8-cm laceration that involves the right cheek, from his right ear to near the corner of his mouth, is noted. Which of the following structures is most likely injured?

A. Parotid duct

B. Submandibular duct

C. Superficial temporal artery

D. Lingual artery

E. Mandibular branch of the facial nerve

45.2 A 45-year-old woman is undergoing surgical resection of the salivary gland for probable cancer. After surgery, she notes that she cannot move her tongue well. Which of the following salivary glands is most likely involved in the surgery?

A. Parotid

B. Sublingual

C. Submandibular

D. Maxillary

45.3 A 16-year-old girl is brought into the physician's office because her mother suspects an eating disorder. The patient has bilateral swelling of the cheeks that are nontender. There are multiple dental caries. She appears to be of normal weight. Which of the following is the most likely diagnosis?

A. Anorexia nervosa

B. Bulimia

C. Irritable bowel syndrome

D. Facetious hyperphagia

ANSWERS

45.1 **A.** The buccal branch of the facial nerve and the parotid duct travel in the area of the cheek and can be located by a line drawn from the tragus of the ear (or the external auditory meatus) to the corner of the mouth.

45.2 **C.** The hypoglossal nerve courses deep to the submandibular gland, and injury to this nerve weakens or paralyzes muscles of the tongue.

45.3 **B.** Enlargement of the parotid glands and multiple dental caries are common in individuals who have bulimia. Affected patients may be of normal weight or even slightly overweight, and their behavior is characterized by binges of eating and inducing vomiting or use of laxatives.

ANATOMY PEARLS

▶ The cranial nerve that passes through the substance of the parotid gland is the facial nerve (CN VII) which innervates the facial muscles. The parotid gland itself receives its parasympathetic secretomotor innervation from the glossopharyngeal nerve (CN IX).

▶ The duct of the parotid gland pierces the buccinator muscle at about the level of the 2nd upper molar tooth, and its termination is usually visible during a physical exam inspection of the cheek mucosa.

▶ During its anteromedial course in the floor of the oral cavity, the submandibular gland's duct is closely related to the hypoglossal and lingual nerves.

▶ The parasympothetic secretomotor innervation of both the submandibular and sublingual glands, is by fibers of the facial nerve (CN VII).

REFERENCES

Gilroy AM, MacPherson BR, Ross LM. *Atlas of Anatomy*, 2nd ed. New York, NY: Thieme Medical Publishers; 2012:580–581.

Moore KL, Dalley AF, Agur AMR. *Clinically Oriented Anatomy*, 7th ed. Baltimore, MD: Lippincott Williams & Wilkins; 2014:914–915, 943–945, 950.

Netter FH. *Atlas of Human Anatomy*, 6th ed. Philadelphia, PA: Saunders; 2014: plates 46, 58–59.

A 43-year-old man was washing his car when he suddenly complained of a severe headache and then slumped to the ground. His son, who witnessed the episode, stated that his father grabbed his head with both hands and cried out in pain as he was falling. The son said that his father had no medical problems and exercised regularly. On examination in the emergency department, the patient was lethargic but responsive to deep pain stimuli. His pupils were dilated bilaterally and sluggishly reactive to light. CT scan of the head showed a significant intracranial hemorrhage. An angiogram demonstrated leakage of dye from the junction of the right internal carotid artery and the circle of Willis.

▶ What is the most likely diagnosis?
▶ What is the clinical anatomy for this event?

ANSWER TO CASE 46:

Berry Aneurysm

Summary: An otherwise healthy 43-year-old man suddenly complained of a severe headache and lost consciousness. He is lethargic, responsive to deep pain, and has bilaterally dilated and sluggishly reactive pupils. CT imaging showed a significant intracranial hemorrhage, and an angiogram demonstrated leakage of dye from the junction of the right internal carotid artery and the circle of Willis.

- **Most likely diagnosis:** Ruptured berry aneurysm

- **Clinical anatomy for this event:** Weakness of the intracranial arterial junction

CLINICAL CORRELATION

This otherwise healthy 43-year-old man had an acute and significant cerebral event. He had a severe headache quickly followed by loss of consciousness. There was no motor activity to suggest an epileptic seizure. Further, his comatose state ruled out self-limited etiologies such as syncope due to a vasovagal reaction. Cerebral imaging confirms intracranial hemorrhage. The possibilities include an arteriovenous malformation (a tangle of vessels that sometimes rupture) or a hemorrhagic stroke. An arteriogram shows leakage of dye from the junction of the internal carotid artery and the circle of Willis, strongly suggesting a berry aneurysm. The blood supply to the brain is derived from the paired internal carotid and the paired vertebral arteries. Occlusion of even one of these vessels would cause severe damage were it not for the anastomosis between these four vessels known as the circle of Willis. However, there is inherent weakness at the junction of the arteries, and an outpouching of the arterial wall, a berry aneurysm, may occur and ultimately rupture. The best treatment for such a ruptured aneurysm is surgical clip ligation. Medications such as calcium channel blockers are useful for preventing coexisting arterial vasospasm.

APPROACH TO:

Vascular Supply of Brain

OBJECTIVES

1. Be able to describe the course of the internal carotid and vertebral arteries

2. Be able to list the major intracranial branches of the internal carotid and the basilar arteries

3. Be able to identify the components of the circle of Willis

DEFINITIONS

ANGIOGRAPHY: Radiographic technique in which contrast medium is injected into the arterial system. X-ray images may be taken at regular intervals to follow the dye from the arterial supply through the venous drainage. Recent advances in MRI have made it possible to examine blood flow without injection of contrast medium.

ANEURYSM: Disruption within the wall of an artery that fills with blood and inflates the muscular coat. The resulting dilation can exert pressure on surrounding structures and ultimately may rupture, leading to a rapid loss of blood pressure.

SYNCOPE: Episode of fainting; a loss of consciousness not related to sleeping.

DISCUSSION

The **arterial blood supply to the brain** is derived from the **paired internal carotid** and **paired vertebral arteries.** The **internal carotid arteries** arise from the **bifurcation of the common carotid arteries** at about the level of the **superior border of the thyroid cartilage.** They are described as being the direct continuation of the common carotids, having no branches in the neck, and ascending to the base of the skull, where they enter the carotid canal. The **internal carotid arteries** pass anteriorly and medially through the **cavernous sinus** to enter the cranial cavity and divide into its terminal branches, the **anterior cerebral artery** and the **middle cerebral artery.** The two anterior cerebral arteries join through the anterior communicating branch. The posterior communicating branch joins the middle cerebral with the posterior cerebral arteries.

The **vertebral arteries** are the first branches of the subclavian arteries in the root of the neck. They ascend through the **transverse foramina of vertebrae C6 through C1**, enter the cranial cavity through the **foramen magnum**, and unite to form the **basilar artery** near the junction of the pons and medulla (Figure 46-1).

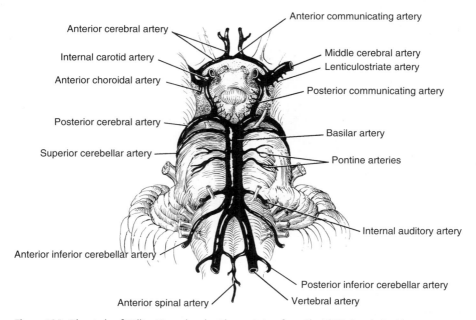

Figure 46-1. The circle of Willis. (*Reproduced, with permission, from Chusid JG. Correlative Neuroanatomy and Functional Neurology, 19th ed. East Norwalk, CT: Appleton & Lange, 1985.*)

At the superior border of the pons, the basilar artery divides into the **posterior cerebral arteries**. The chief intracranial branches of the **vertebral arteries** are the **posterior inferior cerebellar arteries.** Before its terminal bifurcation, the chief branches of the basilar artery are the **anterior inferior cerebellar arteries, superior cerebellar arteries, and several pontine branches.**

The **cerebral arterial circle (of Willis)** is the major anastomosis of the cerebral vasculature. This allows for perfusion of the brain even with arterial occlusion of one or more major arteries (such as carotid insufficiency). If the occlusion develops slowly, the anastomotic vessels will expand to compensate. However, the anastomosis may not be able to compensate if the occlusion develops rapidly, as with trauma. Blockage of one cerebral artery will have characteristic effects based on the region of the brain supplied by the vessel (Figure 46-2). The **anterior cerebral artery** supplies the medial surface of the cerebrum. The **middle cerebral artery** supplies the lateral surfaces, and the **posterior cerebral artery** supplies the inferior surface. The middle

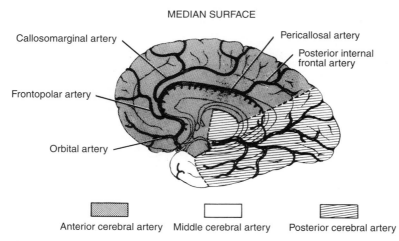

LATERAL SURFACE

Anterior parietal artery

Precentral artery

Central artery

Posterior parietal artery

Orbitofrontal artery

Angular artery

Anterior temporal artery

Posterior temporal artery

MEDIAN SURFACE

Callosomarginal artery

Pericallosal artery

Posterior internal frontal artery

Frontopolar artery

Orbital artery

Anterior cerebral artery   Middle cerebral artery   Posterior cerebral artery

Figure 46-2. Arterial supply of the cerebral cortex. (*Reproduced, with permission, from Lindner HH. Clinical Anatomy. East Norwalk, CT: Appleton & Lange, 1989:32.*)

cerebral artery is occluded most often, presumably because it follows the same trajectory as the internal carotid.

COMPREHENSION QUESTIONS

46.1 A 53-year-old man is being examined for carotid bruits. The physician wants to auscultate the internal carotid artery. At what level does the carotid artery usually bifurcate into the internal and external carotid arteries?

A. At the level of the cricoid cartilage

B. At the level of the thyroid cartilage

C. At the level of sternal notch

D. At the level of the C8 vertebra

46.2 A 64-year-old man is diagnosed with an acute stroke. His main deficit is a partial loss of his visual field. The neurologist diagnoses a lesion of the occipital lobe. Which of the following arteries is likely to be involved?

A. Internal carotid

B. External carotid

C. Middle cerebral

D. Posterior cerebral

46.3 A 35-year-old man complains of the worst headache of his life, grabs onto the back of his neck, and then slumps onto the floor. At the hospital, his CT findings are consistent with a subarachnoid hemorrhage. Which of the following is the most likely etiology?

A. Carotid artery occlusion

B. Vertebrobasilar artery occlusion

C. Middle meningeal artery laceration

D. Rupture of a berry aneurysm

ANSWERS

46.1 **B.** The carotid artery bifurcates at the level of the thyroid cartilage.

46.2 **D.** The occipital lobes are supplied by the posterior cerebral arteries, which are terminal branches of the basilar artery.

46.3 **D.** The most common causes of subarachnoid hemorrhage are rupture of a berry aneurysm in the circle of Willis and bleeding from an arteriovenous malformation.

ANATOMY PEARLS

► The internal carotid arteries have no branches in the neck.

► The terminal branches of the internal carotid arteries are the anterior and middle cerebral arteries.

► The blood supply to the cerebellum is derived from the vertebrobasilar arterial system.

► The circle of Willis allows anastomosis of the arterial blood supply of the brain.

► The major arterial blood supply of the brain is from the internal carotid and vertebral arteries.

REFERENCES

Gilroy AM, MacPherson BR, Ross LM. *Atlas of Anatomy,* 2nd ed. New York, NY: Thieme Medical Publishers; 2012:636–637.

Moore KL, Dalley AF, Agur AMR. *Clinically Oriented Anatomy,* 7th ed. Baltimore, MD: Lippincott Williams & Wilkins; 2014:882–885, 887–888.

Netter FH. *Atlas of Human Anatomy,* 6th ed. Philadelphia, PA: Saunders; 2014: plates 138–142.

A 12-year-old boy complains of a 2-week history of impaired hearing with his left ear. He states that music and voices seem "far away." His medical problems include allergic rhinitis and asthma. On examination, he is afebrile, but his left eardrum displays a yellowish discoloration. The left drum moves very little with a puff of air. The right tympanic membrane appears normal.

▶ What is the most likely diagnosis?
▶ What is the clinical anatomy for this condition?

ANSWER TO CASE 47:

Middle Ear Effusion

Summary: A 12-year-old boy with allergic rhinitis and asthma has a 2-week history of difficulty hearing with his left ear. He is afebrile but has yellowish discoloration of his left tympanic membrane, which does not move well with insufflation.

- **Most likely diagnosis:** Middle ear effusion
- **Clinical anatomy of the condition:** Middle ear fluid impeding sound transmission by the middle ear ossicles

CLINICAL CORRELATION

Sound waves collected by the auricle and external acoustic meatus (canal) produce vibration of the tympanic membrane. These vibrations are transferred, in turn, to the ear ossicles, the malleus, the incus, and the stapes. Vibrations of the stapes produce movements of the endolymph within the cochlea, which are converted to the nervous impulse responsible for the sensation of hearing. Fluid within the middle ear cavity (effusion) diminishes the vibrations of the tympanic membrane and the ear ossicles. Effusions develop in the middle ear secondary to obstruction of the pharyngotympanic (auditory) tube, as with upper respiratory infections or allergic reactions. The insufflation of air through the otoscope in this patient does not induce the normal fluttering of the eardrum, further suggesting an effusion. An infectious process is unlikely in this case because of the absence of a fever or a red eardrum. Treatment of effusions includes antihistamines, decongestants, and, in severe cases, surgical incision of the tympanic membrane for drainage (myringotomy) and insertion of drainage tubes.

APPROACH TO:

The Ear

OBJECTIVES

1. Be able to describe the anatomy of the external acoustic meatus (canal)
2. Be able to describe the anatomy of the tympanic membrane and the three ear ossicles
3. Be able to identify the structures of the middle ear cavity and those that communicate with it

DEFINITIONS

INSUFFLATION: Act of blowing a powder or gas into a body cavity, in this case through the otoscope, to assess whether there is fluid in the middle ear.

PERILYMPH/ENDOLYMPH: The bony labyrinth of the inner ear contains the membranous labyrinth. Within the lumen of the membranous ducts is endolymph, a fluid similar in composition to intracellular fluid (low sodium, high potassium). The space between the ducts and the bony walls is filled with perilymph, a fluid similar in composition to normal extracellular fluid (high sodium, low potassium). The compartments that are filled with perilymph and endolymph do not communicate.

EFFUSION: Spread of a liquid into a space. In this case, the fluid is from the inflammatory response to the allergy.

MYRINGOTOMY: Procedure in which the tympanic membrane is pierced and tubes are inserted into the opening to drain the exudate.

DISCUSSION

The **external ear** is composed of the **auricle**, an elastic cartilage structure covered with skin and having several named parts, one of which is the **concha**, which funnels sound waves down the **external acoustic meatus or canal** (Figure 47-1). The **meatus** is lined with skin, and the wall of the lateral third is cartilaginous, whereas the medial two-thirds are bony. It has an anteromedial S-shaped course, which can be straightened by posterosuperior traction on the auricle.

The medial end of the meatus is closed by the tympanic membrane, a somewhat cone-shaped, 1-cm membrane composed of collagen and elastic fibers that is

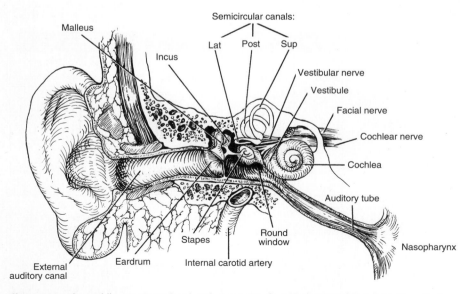

Figure 47-1. The middle ear. (*Reproduced, with permission, from Lindner HH. Clinical Anatomy. East Norwalk, CT: Appleton & Lange, 1989:76.*)

covered externally by thin skin and lined internally by the mucous membrane of the middle ear. The apex of the membrane's cone is called the **umbo.** The reflected light of an otoscope, the **cone of light**, originates at the umbo and is directed **anteroinferiorly.** A process of the **malleus** (also called the "handle") is applied to the medial surface of the membrane, and its tip is also attached at the umbo. The malleus has a lateral process that bulges the superior portion of the membrane laterally. The portion superior to the lateral process is the **pars flaccida**, and the remainder of the membrane is called the **pars tensa.** The three ear ossicles are the **malleus, incus, and stapes**, lateral to medial, across the tympanic or middle ear cavity. Each bone has distinctive features. With a normal tympanic membrane, the handle of the malleus is clearly visible, and the long process of the incus is often visible posterior to the malleus. The **stapes** is shaped much like a **stirrup**, and its footplate fits into the oval window on the medial wall of the tympanic cavity. Its in-and-out movement transmits pressure waves through the **endolymph** within the **cochlea**, where the nerve impulses for hearing are generated. Excessive movements of the ear ossicles with loud noise are dampened by the tensor tympani muscle, which is attached to the malleus, and the stapedius muscle, which is attached to the stapes. These **muscles are innervated by CN V and CN VII, respectively.**

The **tympanic cavity** is contained within the **petrous portion of the temporal bone.** Its features are usually described as being contained within a box with a roof, four walls, and a floor. Table 47-1 lists the bony features, related structures, and openings for each of the walls. The tympanic cavity is lined with a mucous membrane and contains the **chorda tympani branch of CN VII and the tympanic plexus (CN IX)** in addition to the ear ossicles and their associated muscles. Air pressure within the cavity is equalized with the nasopharynx through the pharyngotympanic or auditory tube.

Table 47-1 • WALLS OF THE TYMPANIC CAVITY						
	Roof	**Floor**	**Lateral Wall**	**Medial Wall**	**Anterior Wall**	**Posterior Wall**
Bony Feature	Tegmen tympani of temporal bone		Tympanic membrane, malleus, epitympanic recess	Promontory, prominence of facial canal, prominence of lateral semicircular canal		Mastoid process, pyramid
Related Structure	Middle cranial fossa	Internal jugular vein	External acoustic meatus, chorda tympani	Vestibular apparatus, CN VII	Carotid artery, tensor tympani muscle	Mastoid air cells, CN VII, stapedius muscle
Opening					Auditory tube	Mastoid aditus

COMPREHENSION QUESTIONS

47.1 A 4-year-old boy was noted to have recurrent ear infections. He underwent placement of tubes in the tympanic membranes 3 days previously and currently complains of some difficulty in tasting candy. Which of the following is the most likely explanation?

    A. Disruption of CN VIII

    B. Disruption of the chorda tympani

    C. Effects of the anesthesia

    D. Effects of the endotracheal tube

47.2 A 5-year-old girl complains of severe pain from her right ear due to an acute otitis media. Which of the following nerves is most likely responsible for carrying the sensation of pain from the tympanic membrane?

    A. CN VII

    B. CN VIII

    C. CN IX

    D. CN X

47.3 A 3-year-old boy had three episodes of otitis media over the past year. His mother asks the pediatrician why children tend to develop more ear infections than adults. Which of the following is the most likely anatomical explanation?

    A. Changes in the eustachian tube

    B. Changes in the external pinna

    C. Changes in the external ear canal

    D. Changes in the stapedius ossicle

ANSWERS

47.1 **B.** The chorda tympani, which is a branch of CN VII, courses behind the tympanic membrane and occasionally can be injured during surgery for ear tubes. The chorda tympani innervates the anterior two-thirds of the tongue.

47.2 **C.** The glossopharyngeal nerve (CN IX) is the afferent nerve for the sensory input from the internal surface of the tympanic membrane and the tympanic cavity.

47.3 **A.** The eustachian tube connects the middle ear to the oral cavity. The eustachian tube is shorter and more horizontal in a child than in an adult.

ANATOMY PEARLS

▶ The outer one-third of the external acoustic meatus is cartilaginous, thereby facilitating the straightening of its S-shaped curvature.

▶ The cone of light is seen in the anteroinferior quadrant of the tympanic membrane.

▶ The tensor tympani and stapedius muscles of the middle ear are innervated by CN V and CN VII, respectively.

▶ The tympanic cavity communicates with the nasopharynx through the pharyngotympanic (auditory) tube.

REFERENCES

Gilroy AM, MacPherson BR, Ross LM. *Atlas of Anatomy*, 2nd ed. New York, NY: Thieme Medical Publishers; 2012:560–565.

Moore KL, Dalley AF, Agur AMR. *Clinically Oriented Anatomy*, 7th ed. Baltimore, MD: Lippincott Williams & Wilkins; 2014:967–973, 978–979.

Netter FH. *Atlas of Human Anatomy*, 6th ed. Philadelphia, PA: Saunders; 2014: plates 94–96, 98, 100.

A 10-year-old girl is brought to her pediatrician's office complaining of headache for the past 2 weeks. Her mother had taken the girl to an optometrist, and her vision was normal. The patient states that she has been in good health and that she received a cat as a birthday present 1 month previously. On examination, she has a normal temperature, the tympanic membranes appear normal, and her throat is clear. There is some tenderness of the right cheek and over the right orbit.

▶ What is the most likely diagnosis?
▶ What is the anatomical explanation for this condition?

ANSWER TO CASE 48:

Sinusitis

Summary: A 10-year-old girl who had recently acquired a cat has had a headache for a 2-week duration. She is afebrile, with normal-appearing tympanic membranes and throat. She has right maxillary and frontal tenderness.

- **Most likely diagnosis:** Maxillary and frontal sinusitis

- **Anatomical explanation for this condition:** Blocked drainage of the sinuses secondary to an allergic reaction of the nasal mucosa

CLINICAL CORRELATION

Sinusitis, a condition common in Americans, is an inflammation of one or more of the six sets of paranasal sinuses, most of which are related to the orbits. Inflammation may be caused by viruses, allergies, and bacterial pathogens. The sinuses are usually sterile cavities that are lined by ciliated mucosa rich in mucous cells, and mucus drains directly into the nasal cavities through small openings, or ostia. Edema of the nasal mucosa can easily occlude these openings and lead to secondary infection. The maxillary sinus is most commonly involved, and sinus pain or pressure sensation is typical. Transillumination of the sinuses that demonstrates opacification may be helpful on physical examination. Radiographs may also be helpful; CT imaging is usually reserved for complicated cases. The recent acquisition of a cat by the patient suggests maxillary and frontal sinusitis caused by an allergy rather than an infectious agent. Oral or topical (spray) decongestants, antihistamines, and/or nasal steroids are often helpful. Antibiotics are not indicated at this time, but the patient should be instructed to watch for development of fever or an increase in tenderness. Complications include osteomyelitis, ocular cellulitis, and cavernous sinus thrombophlebitis.

APPROACH TO:

The Sinuses

OBJECTIVES

1. Be able to describe the location of the paranasal sinuses in the facial skeleton

2. Be able to list the openings in the nasal cavity through which the paranasal sinuses drain

DEFINITIONS

OSTEOMYELITIS: Condition in which the bone and bone marrow become infected

THROMBOPHLEBITIS: Condition in which a vein becomes inflamed in response to a blood clot

DISCUSSION

The **paranasal sinuses are extensions of the nasal cavities into bones of the skull** and are **named for the bones in which they are located** (Figure 48-1). These spaces are lined with respiratory mucosa, decrease the weight of the skull, and probably assist in humidifying inspired air. See Case 51 for the anatomy of the nasal cavity. The **sphenoid sinuses** are located within the sphenoid bone, are variable in size and number, and open into the sphenoethmoidal recess. The **ethmoidal sinuses** consist of a series of sinuses positioned between the medial wall of the orbit and the nasal cavity (at the level of the bridge of the nose). For descriptive purposes, they are divided into anterior, middle, and posterior ethmoidal cells, and each has a separate opening. The **posterior ethmoidal cells open inside the superior nasal meatus.** The **middle ethmoidal cells** elevate the ethmoid bone in the middle meatus, thus creating the ethmoid bulla on whose surface these cells open. Inferior to the ethmoid bulla is a groove, the semilunar hiatus. The **anterior ethmoidal cells** open into the anterior portion of the hiatus, called the **infundibulum.**

The **largest sinuses are the maxillary and frontal sinuses**, and their relatively large openings also drain into the **middle meatus.** The large **maxillary sinus** hollows the maxillary bone. The roof of the sinus, which also forms the floor of the orbit, is very thin and at risk in direct trauma to the orbit, which would cause sudden increases in pressure. Such trauma may cause "blowout" fractures of the orbital floor. The opening of the maxillary sinus is found in the semilunar hiatus. The **frontal sinuses** are found in the frontal bone between the inner and outer tables and in the portion that forms the roof of the orbit. It is drained by the **frontonasal duct**, which opens into the infundibulum, the anterior portion of the semilunar hiatus.

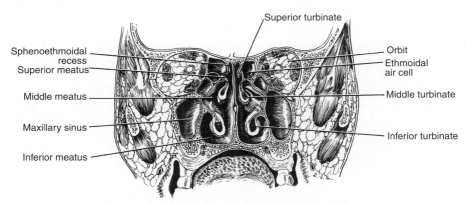

Figure 48-1. Sinuses in the coronal view. (*Reproduced, with permission, from Lindner HH. Clinical Anatomy. East Norwalk, CT: Appleton & Lange, 1989:68.*)

COMPREHENSION QUESTIONS

48.1 A 24-year-old medical student has been diagnosed with sinusitis and asks her physician why there is nasal drainage during the night but not during the day. Which of the following is the best explanation?

A. Location of the ostia within the sinus

B. Location of the ostia within the nasal passage

C. Disruption of the drainage due to mastication

D. Diurnal mucus production increases at night

48.2 A 22-year-old college student is being seen for possible sinusitis. The physician sees purulent drainage arising from the superior nasal meatus. Which of the following sinuses is likely to be infected?

A. Frontal

B. Maxillary

C. Sphenoidal

D. Ethmoidal

48.3 A 28-year-old neuroanatomy graduate student noted pain at the bridge of his nose and had been told that he had "sinus" infections. He was speculating about the afferent nerve supply from this area. Which of the following is the most accurate description of the sensory nerve innervation?

A. Branches of CN III

B. Branches of CN V

C. Branches of CN VII

D. Branches of CN IX

ANSWERS

48.1 **A.** The sinus most likely affected is the maxillary sinus. The ostia within the sinus are located superiorly in a location inefficient for gravity drainage. During sleep at night, the mucus flows out through the ostia.

48.2 **D.** The posterior ethmoidal sinus drains into the superior nasal meatus.

48.3 **B.** The paranasal sinuses are innervated by branches of CN V.

ANATOMY PEARLS

▶ The paranasal sinuses are named after the bones in which they are found (frontal, ethmoid, sphenoid, and maxilla).

▶ The maxillary, frontal, anterior, and posterior ethmoidal sinuses open into the middle nasal meatus.

▶ The maxillary sinus is the largest of the paranasal sinuses and is the most commonly infected sinus because its ostia are located superiorly.

▶ Trauma to the orbit may result in a blowout fracture and, hence, orbital structures (such as extraocular muscles) may be pushed inferiorly into the maxillary sinus.

REFERENCES

Gilroy AM, MacPherson BR, Ross LM. *Atlas of Anatomy,* 2nd ed. New York, NY: Thieme Medical Publishers; 2012:537, 552–553.

Moore KL, Dalley AF, Agur AMR. *Clinically Oriented Anatomy,* 7th ed. Baltimore, MD: Lippincott Williams & Wilkins; 2014:960–964.

Netter FH. *Atlas of Human Anatomy,* 6th ed. Philadelphia, PA: Saunders; 2014: plates 42–45.

A 7-year-old boy was referred to an ear-nose-throat (ENT) specialist after experiencing recurrent episodes of tonsillitis with peritonsillar abscesses. His mother noted about seven infections over the past 8 months, all treated with antibiotics. After discussing treatment options with the family, the ENT specialist recommended a tonsillectomy. The patient's tonsillectomy was complicated by bleeding from the surgical bed, and he had temporary loss of taste sensation from his posterior tongue. He is currently doing well and without complaints.

► The intraoperative bleeding was most likely from which blood vessel?
► Why was there a temporary loss of taste sensation?

ANSWER TO CASE 49:

Recurrent Tonsillitis

Summary: A 7-year-old boy is status posttonsillectomy for recurrent tonsillitis complicated by increased intraoperative bleeding and temporary loss of taste sensation from the posterior one-third of the tongue.

- **Vessel involved with intraoperative bleeding:** External palatine vein

- **Loss of taste sensation:** Compression of glossopharyngeal nerve (CN IX)

CLINICAL CORRELATION

For patients with recurrent episodes of tonsillitis or peritonsillar abscesses (>4 episodes per year), tonsillectomy may be indicated. Although tonsillectomy is regarded as a routine procedure, it is not without complications and risks. A thorough understanding of the anatomy of the pharynx is necessary in order to limit complications. The tonsillar bed is extremely vascular with the most common source of intraoperative bleeding from the external palatine vein, which arises from the lateral aspect of the tonsillar bed. Even if a direct injury does not occur, compression from edema may cause temporary injury as in this case. Compression of glossopharyngeal nerve branches causes a temporary loss in taste sensation on the posterior aspect of the tongue. As the swelling decreases, so does the nerve impairment. Many other vital vessels, nerves, and structures are adjacent to the tonsils, and care must be taken to avoid injury.

APPROACH TO:

The Tonsils

OBJECTIVES

1. Be able to describe the divisions of the pharynx

2. Be able to list the muscles that form the pharynx

3. Be able to describe the components of the tonsillar ring

4. Be able to identify vessels that supply the pharynx, especially branches that course through the tonsillar beds

5. Be able to identify the cranial nerves providing sensory and motor innervation to the pharynx

DISCUSSION

The **pharynx** is a space within the head that connects the **oral and nasal cavities** to the **trachea** and **esophagus**. Air-filled spaces in the temporal bone (i.e., the tympanic cavity and the mastoid air cells) connect with the pharynx through the **pharyngotympanic (eustachian) tube.** The walls of the pharynx are covered with mucosa. Deep to the mucosa are several aggregations of lymphoid tissue that form a ring around the pharynx, priming the immune system for defense against pathogens (see Figure 49-1).

The superior boundary of the pharynx is the base of the skull. The muscles of the pharyngeal walls form a cone that narrows to the esophagus. The medial pterygoid plates support the lateral walls of the superior part of the pharynx. The bodies of cervical vertebrae support the posterior wall. The anterior wall is interrupted by three apertures. One opens to the nasal cavity, another to the oral cavity, and a third to the larynx. Therefore, the pharynx is divided into three corresponding regions: the **nasopharynx**, the **oropharynx**, and the **laryngopharynx.** The naso- and oropharynx are continuous but are separated by elevation of the soft palate during swallowing to prevent reflux of food and liquid into the nasopharynx. The oro- and laryngopharynx are also continuous. Depression of the **epiglottis** during swallowing separates the larynx from the laryngopharynx, preventing aspiration into the trachea and lungs.

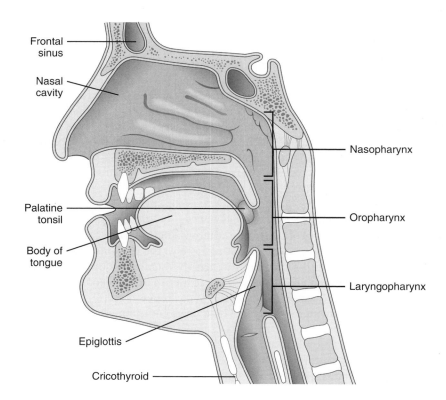

Figure 49-1. Median section through pharynx.

The pharyngeal wall is composed of three muscles: the **superior, middle, and inferior pharyngeal constrictors.** The inferior part of the inferior constrictor muscle thickens as it merges with the esophagus, forming a sphincter called the **cricopharyngeus muscle.** The three constrictor muscles are stacked like ice cream cones. Between the pairs of muscles are gaps that transmit important structures. The gap between the superior constrictor and the occipital bone transmits the pharyngotympanic tube, the **levator veli palatini muscle,** and the **ascending palatine artery.** Between the superior and middle constrictors are the glossopharyngeal nerve and **stylopharyngeus muscle.** Between the middle and inferior constrictors course the **internal laryngeal nerve** and the **superior laryngeal artery.** The **recurrent laryngeal nerve** and the **inferior laryngeal artery** ascend deep to the inferior constrictor.

The lymphoid tissue surrounding the pharynx is commonly called the **Waldeyer ring,** which is composed of three masses of lymphoid tissue: the **pharyngeal tonsils** (also called "adenoids" when enlarged), the **palatine tonsils,** and the **lingual tonsils.** The pharyngeal tonsils are located in the roof and posterior wall of the nasopharynx. The opening of the pharyngotympanic tube into the nasopharynx is protected by a tonsil. The palatine tonsils are located in the anterior wall of the oropharynx between the **palatoglossal and palatopharyngeal folds.** The lingual tonsil is located under the mucosa of the posterior one-third of the tongue.

The pharynx is supplied by arteries from several sources, most of which are branches of the external carotid artery, specifically the **maxillary, facial, lingual, and superior thyroid arteries.** The constrictor muscles are also supplied by branches from the **deep cervical and inferior thyroid arteries.** With respect to this case, the most important vessels are the ascending palatine and tonsillar branches of the facial artery. Surgery to remove the palatine tonsil can damage the tonsillar branch, resulting in excessive bleeding. Venous drainage from the pharynx parallels the arterial supply. In addition, there is an extensive **pharyngeal venous plexus** on the posterior surface of the constrictor muscles. The **external palatine vein** descends along the lateral surface of the palatine tonsil to drain into the venous plexus. Therefore, this vessel may be damaged during surgery to remove a palatine tonsil, also resulting in excessive bleeding.

The nerve supply to the pharynx is from cranial nerves IX and X. The **glossopharyngeal nerve (CN IX)** supplies general sensory fibers to the mucosa of the pharynx. These fibers contribute to the **afferent limb of the gag reflex.** CN IX also supplies special sensory fibers mediating taste to the posterior one-third of the tongue. This nerve exits the cranium through the **jugular foramen** and descends with the **stylopharyngeus muscle** to pass through the gap between the superior and middle pharyngeal constrictor muscles. The vagus nerve (CN X) supplies general motor fibers to the constrictor muscles. These fibers contribute to the **efferent limb of the gag reflex.** This nerve also exits the cranium through the jugular foramen but descends within the carotid sheath. As it descends, it gives off branches that form the pharyngeal plexus on the posterior surface of the pharynx. In this case, edema from the tonsillectomy compressed the branches of CN IX, blocking the sensation of taste from the posterior one-third of the tongue.

COMPREHENSION QUESTIONS

49.1 During a procedure to remove a palatine tonsil, the operating field was suddenly filled with bright red blood. Which artery was inadvertently damaged?

A. Tonsillar branch of facial

B. Ascending pharyngeal

C. Ascending palatine

D. Descending palatine

E. Lingual

49.2 A patient has a mild chronic cough but has clear lungs and no evidence of bronchitis. Her physician believes that the symptoms are due to postnasal drip brought on by allergy. Which nerve is responsible for the afferent limb of the cough reflex?

A. CN V2

B. CN V3

C. CN VII

D. CN IX

E. CN X

49.3 Which structure passes through the gap between the superior and middle constrictor muscles?

A. Recurrent laryngeal artery

B. Internal laryngeal nerve

C. Superior laryngeal artery

D. Glossopharyngeal nerve

E. Pharyngotympanic tube

ANSWERS

49.1 **A.** The tonsillar branch of the facial artery lies in the bed of the palatine tonsil and is susceptible to injury. Although the ascending palatine artery sends branches to the tonsil, it is unlikely to be affected in a routine procedure.

49.2 **D.** The cough reflex is stimulated by irritation of the laryngopharynx, which is innervated by CN IX. The trigeminal nerve (CN V1 and V2) innervates the oral and nasal cavities.

49.3 **D.** The glossopharyngeal nerve (CN IX) passes through the gap between the superior and middle constrictors, along with the stylopharyngeus muscle and stylohyoid ligament.

ANATOMY PEARLS

▶ The three pharyngeal constrictor muscles are stacked like ice cream cones. Structures pass into the pharynx through gaps between the muscles.

▶ The tonsillar (Waldeyer) ring is a discontinuous mass of lymphoid tissue located where the body opens to the environment, exposing the immune system to pathogens.

▶ At the base of the palatine tonsil, the tonsillar branch of the facial artery and the glossopharyngeal nerve (CN IX) can be identified.

▶ The gag reflex is evoked by mechanical stimulation of the oropharynx. The afferent limb of the reflex is mediated by the glossopharyngeal nerve (CN IX), and the efferent limb is mediated by the vagus nerve (CN X).

REFERENCES

Gilroy AM, MacPherson BR, Ross LM. *Atlas of Anatomy*, 2nd ed. New York, NY: Thieme Medical Publishers; 2012:582–583, 586–587.

Moore KL, Dalley AF, Agur AMR. *Clinically Oriented Anatomy*, 7th ed. Baltimore, MD: Lippincott Williams & Wilkins; 2014:1032–1036, 1047–1048.

Netter FH. *Atlas of Human Anatomy*, 6th ed. Philadelphia, PA: Saunders; 2014: plates 64, 68.

A 47-year-old woman is undergoing surgical removal of her gallbladder (chole-cystectomy). Her medical problems include insulin-dependent diabetes melli-tus and sleep apnea. After the anesthesiologist has administered the paralyzing agent (succinylcholine), the patient experiences spasms of the airway and diffi-culty breathing with the bag and mask. The anesthesiologist attempts to place an endotracheal tube by direct visualization (direct laryngoscopy), without success due to swelling (laryngeal edema). Meanwhile, the oxygen saturation content of the blood has decreased to a very low range of 80 percent. The anesthesiologist remarks that an emergency airway needs to be surgically opened.

► What is your next step?
► What anatomical landmarks will be most helpful?

ANSWER TO CASE 50:

Emergency Tracheostomy

Summary: A 47-year-old woman with a history of diabetes and sleep apnea is undergoing elective cholecystectomy. After receiving the paralyzing agent, the patient develops laryngospasm and is difficult to ventilate. Direct laryngoscopy and intubation attempts are unsuccessful, and oxygen saturation is low.

- **Next step:** Emergency tracheostomy or cricothyroidotomy
- **Helpful anatomical landmarks:** Cricoid and thyroid laryngeal cartilages

CLINICAL CORRELATION

A leading cause of mortality at elective surgery is related to anesthesia, specifically an inability to ventilate the patient. This woman is probably obese and difficult to intubate because of her short neck, and her sleep apnea is a concern. When oxygen saturation decreases to dangerous levels (<90 percent), brain and/or heart ischemia may ensue. Immediate correction of oxygenation is critical, and, as in this case, emergency tracheostomy is indicated. One of the most expedient methods is to enter the cricothyroid membrane in the midline, between the cricoid and thyroid laryngeal cartilages. This interval is usually palpable and is approximately one-third the distance from the top of the manubrium to the tip of the chin (mentum). A vertical incision is made in the membrane, and a tracheal tube is inserted. Alternatively, a needle can be inserted into the same membrane, and oxygen can be administered through a jet ventilator. However, this procedure must be revised rapidly because there is insufficient flow to remove carbon dioxide from the lungs. Nonemergency tracheostomies are performed inferiorly to the cricoid cartilage and the isthmus of the thyroid gland.

APPROACH TO:

The Neck: Upper Airway

OBJECTIVES

1. Be able to list the landmarks of the anterior neck and identify the muscles of the infrahyoid region
2. Be able to describe the cartilaginous skeleton of the larynx and the positions of the vocal cords in relation to palpable landmarks
3. Be able to describe the thyroid gland's relationship to the larynx and its blood supply

DEFINITIONS

ABCs: This mnemonic reminds us that the priorities of emergency management are airway, breathing, and circulation.

ENDOTRACHEAL INTUBATION: Placement of a tube through the mouth or nose and through the vocal cords to secure the airway and/or provide mechanical ventilation.

TRACHEOSTOMY: Surgical establishment of an airway by an opening from the skin to the trachea. These are emergent when endotracheal intubation is impossible, and they are elective when the patient has need of a long-term airway.

CRICOTHYROIDOTOMY: Temporary method of establishing an airway by penetrating through the cricothyroid membrane. The procedure is nearly always performed emergently.

CHOLECYSTECTOMY: Surgical procedure to remove the gallbladder.

SLEEP APNEA: Condition in which the patient in unable to breathe because of temporary obstruction of the airway, usually occurring during sleep. Loud snoring, choking, or periods of cessation of breathing are typical.

DISCUSSION

Deep to the thin skin of the anterior neck is the **platysma muscle**, which is within the **superficial fascia.** Deep to the platysma are the **infrahyoid ("strap") muscles** of the neck. The paired **sternohyoid muscles** extend from the posterior surface of the manubrium to the hyoid bone, and their medial borders are just lateral and parallel to the midline. The superior bellies of the **omohyoid muscles** lie just lateral to the sternohyoid muscles. Deep to these muscles are found the **sternothyroid muscles,** and continuing superiorly are the **thyrohyoid muscles.**

The skeleton of the **larynx** consists of the U-shaped **hyoid bone**, which lies at the level of the **C3 vertebra**, and nine cartilages. The **epiglottis, thyroid, and cricoid cartilages** are unpaired, whereas the **arytenoids, corniculate, and cuneiform are paired.** The thyroid cartilage, which resembles an open book, lies opposite the C4 and C5 vertebrae. Its two laminae are united anteriorly, and the **laryngeal prominence (Adam's apple)** is easily palpated and typically visible in men. The **cricoid cartilage is shaped like a signet ring;** its larger laminar portion is posterior. It lies opposite the **C6 vertebra.** The thyroid cartilage is joined to the hyoid bone above and the cricoid cartilage below by ligaments and membranes. The **true vocal cords extend from the vocal processes of the arytenoid cartilages atop the lamina of the cricoid cartilage to the posterior surface of the thyroid cartilage superior to the lower border of the cartilage** (Figure 50-1). The interval between the thyroid and cricoid cartilages is closed by the cricothyroid membrane and is inferior to the true vocal cords (Figure 49-1). The cricothyroid muscle is also found laterally in this interval.

The **thyroid gland, like the larynx, is enclosed within the pretracheal fascia.** The large laterally placed lobes of the gland are applied to the surface of the laminae of the thyroid cartilage and the upper trachea, with the **parathyroid glands**

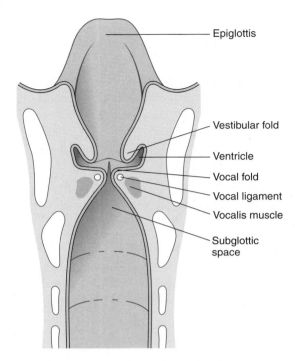

Epiglottis

Vestibular fold

Ventricle

Vocal fold

Vocal ligament

Vocalis muscle

Subglottic space

Figure 50-1. The larynx in coronal section, including the vocal cords.

embedded in their posterior surfaces. The right and left lobes are joined across the midline by the **isthmus**, which typically is inferior to the cricoid cartilage at the level of the second and third tracheal cartilage rings. In approximately 50 percent of individuals, a pyramidal lobe may be present that extends superiorly and overlies the cricothyroid membrane, but usually to one side of the midline. This **remnant of the thyroglossal duct** may be glandular or fibrous tissue. The thyroid and parathyroid glands are supplied by the paired **superior thyroid arteries** (direct branches from the **external carotid arteries**) and the **inferior thyroid arteries**, which are branches from the thyrocervical trunk. In 12 percent of individuals, a **small midline artery, the thyroid ima artery**, arises directly from the aortic arch or brachiocephalic trunk. It ascends on the anterior surface of the trachea to reach the isthmus.

COMPREHENSION QUESTIONS

50.1 A 24-year-old man is being evaluated for airway abnormalities. Palpation of the cricoid cartilage is normally at which vertebral level?

A. C2

B. C4

C. C6

D. T1

50.2 A 45-year-old woman is undergoing thyroid surgery for suspected thyroid cancer. The surgeon has taken a midline approach and encounters significant bleeding below the isthmus of the thyroid gland. Which of the following is the likely cause of the bleeding?

   A. Penetration into the trachea
   B. Superior thyroid artery
   C. Inferior thyroid artery
   D. Thyroid ima artery
   E. Inferior laryngeal artery

50.3 A 54-year-old woman has undergone partial thyroid resection due to a nontender cold nodule that likely represents cancer. One week after surgery, she complains of twitching of the right arm and "spasms" of both hands. Which of the following is the most likely explanation?

   A. Anxiety after surgery
   B. Effects of anesthesia
   C. Parathyroid glands removed
   D. Vagal nerve injury

50.4 An emergency cricothyroidotomy is thought to be warranted because of airway collapse and severe laryngoedema. Which of the following is the most accurate description of the location of the cricothyroid membrane?

   A. Immediately superior to the thyroid cartilage
   B. Immediately inferior to the thyroid cartilage
   C. Immediately inferior to the cricoid cartilage
   D. Just deep to the isthmus of the thyroid gland
   E. Immediately inferior to the hyoid bone

ANSWERS

50.1 **C.** The cricoid cartilage is usually located at the C6 vertebral level.

50.2 **D.** In up to 12 percent of individuals, a small midline artery, called the thyroid ima artery, arises from the aortic arch or brachiocephalic trunk and reaches the thyroid isthmus inferiorly.

50.3 **C.** The parathyroid glands are variably inside the thyroid gland. With resections of the thyroid, the small parathyroid glands may be affected, leading to decreased levels of calcitonin and, hence, hypocalcemia. The low calcium levels may cause clinical symptoms of muscle spasms, tetany, or even convulsions.

50.4 **B.** The cricothyroid membrane is just inferior to the thyroid cartilage and superior to the cricoid cartilage.

ANATOMY PEARLS

▶ The cricoid cartilage lies at the level of the C6 vertebra.

▶ The true vocal cords lie superior to the cricothyroid membrane.

▶ The cricothyroid membrane is located inferior to the thyroid cartilage and superior to the cricoid cartilage.

▶ A pyramidal thyroid lobe may overlay the cricothyroid membrane, close to the midline.

▶ In a small percentage of patients, a small midline artery, the thyroid ima, may directly supply the isthmus.

REFERENCES

Gilroy AM, MacPherson BR, Ross LM. *Atlas of Anatomy*, 2nd ed. New York, NY: Thieme Medical Publishers; 2012:598–599, 601, 603.

Moore KL, Dalley AF, Agur AMR. *Clinically Oriented Anatomy*, 7th ed. Baltimore, MD: Lippincott Williams & Wilkins; 2014:1018–1029, 1030–1032, 1045.

Netter FH. *Atlas of Human Anatomy*, 6th ed. Philadelphia, PA: Saunders; 2014: plates 70, 76, 79–80.

A 22-year-old male presents to the emergency department complaining of severe, unstoppable nasal bleeding for the past 30 min. He denies any trauma, bleeding disorders, or use of medications such as aspirin or ibuprofen. The patient indicates that this nosebleed is unique because he is bleeding from both nostrils and blood is draining into his throat and choking him. He feels as though the blood were collecting in the back of his throat. He has tried pinching his nose, but the bleeding continues.

▶ What is the most likely anatomical explanation for this condition?

ANSWER TO CASE 51:

Epistaxis

Summary: A 22-year-old male has had a 30-min bilateral epistaxis with drainage of blood to the nasopharynx and choking. He denies trauma, bleeding disorders, or use of anticoagulant medications. Anterior nasal pinching did not help.

- **Most likely anatomical explanation:** Posterior epistaxis

CLINICAL CORRELATION

Epistaxis, or bleeding from the nose, is a common condition. Most cases arise from the anterior region of the nasal septum, and the bleeding site is fairly easy to visualize. Most anterior nosebleeds will respond to direct pressure, although other measures may be necessary, including topical vasoconstrictors such as cocaine, cautery, or nasal packing. This patient's epistaxis is atypical in that it is bilateral, with posterior drainage that produces a choking sensation. These symptoms suggest a posterior source, which is more difficult to control. Treatment of this type is by posterior nasal pack or a balloon tamponade device. Antibiotics are usually required to prevent sinusitis or toxic shock syndrome. Persistent or atypical epistaxis should alert the clinician to bleeding abnormalities. Patients who have congenital conditions such as hemophilia or von Willebrand disease may develop epistaxis. Acquired processes, such as use of aspirin or nonsteroidal anti-inflammatory medication, or frank anticoagulation with heparin or warfarin sodium (Coumadin) may be causative. Disease processes such as hepatic failure may lead to decreased levels of vitamin K–dependent coagulation factors.

APPROACH TO:

The Nose

OBJECTIVES

1. Be able to list the features of the external nose and nasal cavity

2. Be able to describe the arterial supply to the nasal cavities

DEFINITIONS

EPISTAXIS: Bleeding from the nose, usually divided clinically into an anterior or a posterior source.

KIESSELBACH PLEXUS: Area on the anterior portion of the nasal septum that is very vascular because of the anastomosis of blood vessels; this is the most common site for epistaxis.

COAGULOPATHY: Abnormalities to the normal pathways of hemostasis that lead to bleeding. Causes are usually congenital or acquired.

ANTICOAGULANT: Chemical that interferes with the normal process of blood clotting.

DISCUSSION

The **external nose** is composed of the paired nasal bones, which form the bridge of the nose, and adjacent portions of the frontal bones and maxillae. The majority of the external nose is **cartilaginous** and is formed by the paired **alar** and **lateral nasal cartilages** and the **unpaired septal cartilage.** The anterior opening into the nasal cavity is the **anterior nares.** The nasal cavity is a somewhat pyramidal space within the skull located between the two orbits. It is subdivided into right and left nasal cavities by the **nasal septum,** which is formed by the **vomer bone, perpendicular plate** of the **ethmoid bone,** nasal crests of the maxilla and palatine bones, and the **septal cartilage.** The roof of each cavity is formed by the **frontal, ethmoid, and sphenoid bones,** and its floor is formed by the palatine portion of the maxilla and the **horizontal plate of the palatine bone.** The posterior openings of each nasal cavity into the nasopharynx are the posterior **choanae.**

The complex lateral walls are formed by portions of the nasal, maxilla, ethmoid, and palatine bones. The surface area of the lateral walls is increased by the three nasal **conchae.** The **superior and middle conchae** are features of the **ethmoid bone,** whereas the **inferior nasal concha is an individual bone.** The posterosuperior portion of the nasal cavity, superior to the superior conchae, is the **sphenoethmoid recess.** Inferior to each of the conchae is a space named for the concha immediately superior to it. Thus, the superior, middle, and inferior nasal meatuses lie inferiorly to the superior, middle, and inferior nasal conchae, respectively. Each nasal cavity is lined with a highly vascular mucosa whose function is to warm and humidify inspired air (Figure 51-1).

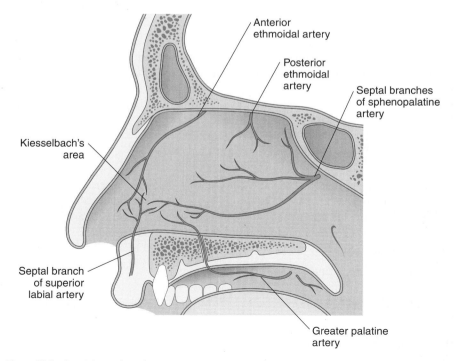

Anterior ethmoidal artery

Posterior ethmoidal artery

Septal branches of sphenopalatine artery

Kiesselbach's area

Septal branch of superior labial artery

Greater palatine artery

Figure 51-1. Arterial supply to the nose (septum).

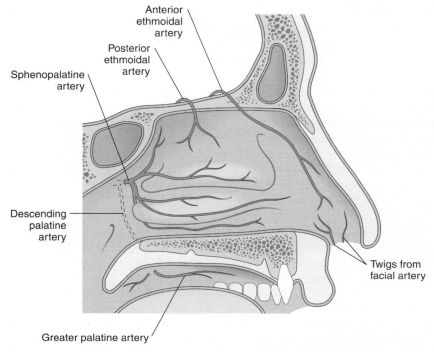

Figure 51-2. Arterial supply to the nose (lateral wall).

Each nasal cavity is supplied by nasal branches of the **sphenopalatine artery, anterior and posterior ethmoidal arteries, greater palatine artery, and superior labial and lateral nasal branches of the facial artery** (Figure 51-2). These arteries anastomose at **Kiesselbach area** on the anterior portion of the nasal septum (opposite the anterior end of the inferior concha). This is the **most common site for epistaxis.**

COMPREHENSION QUESTIONS

51.1 A 55-year-old man has become anemic and hypotensive due to severe anterior epistaxis. An ear-nose-throat (ENT) surgeon has been called to address the bleeding. He states that he may need to occlude the major arterial supply. Which of the following arteries is most likely to be responsible?

A. Ethmoidal

B. Sphenopalatine

C. Superior labial

D. Greater palatine

51.2 An 18-year-old woman arrives in the emergency department complaining of persistent epistaxis. On examination, there is bleeding from the right nostril. Which of the following locations is the most likely source of the bleeding?

A. Anterior nasal septum

B. Posterior nasal septum

C. Anterior turbinate

D. Posterior turbinate

E. Nasal floor

51.3 A 24-year-old woman is thrown from her car during a motor vehicle accident and hits her head against the pavement. She has lost consciousness but currently is alert and has equally reactive pupils. She is asymptomatic except for clear nasal leakage from the right nostril that has not abated over 24 h. Which of the following is the most likely etiology?

A. Sympathetic sinus drainage

B. Allergic rhinitis from the car's airbag

C. Damage to the cribriform plate

D. Lacrimonasal fistula

ANSWERS

51.1 **B.** The major blood supply to the anterior septum is the sphenopalatine artery, a branch of which supplies the nasal septum. The sphenopalatine artery arises from the maxillary artery, which is a terminal branch of the external carotid artery.

51.2 **A.** The most common location of epistaxis is the region of the anterior septum known as **Kiesselbach plexus**, which has a rich anastomosis of arteries.

51.3 **C.** This patient likely has cerebrospinal fluid (CSF) rhinorrhea, which is not unusual following head trauma. The cribriform plate and meninges are disrupted, thus allowing CSF to leak through the nose. This predisposes one to meningitis.

ANATOMY PEARLS

► The anterior portion of the nasal septum is cartilaginous.

► The superior and middle nasal conchae are features of the ethmoid bone.

► The most common site for epistaxis is where the several arteries that supply the nasal cavity anastomose on the anterior nasal septum (Kiesselbach area).

REFERENCES

Gilroy AM, MacPherson BR, Ross LM. *Atlas of Anatomy,* 2nd ed. New York: Thieme Medical Publishers; 2012:554–555.

Moore KL, Dalley AF, Agur AMR. *Clinically Oriented Anatomy,* 7th ed. Baltimore, MD: Lippincott Williams & Wilkins; 2014:959–960, 964.

Netter FH. *Atlas of Human Anatomy,* 6th ed. Philadelphia, PA: Saunders; 2014: plates 36–40.

A 45-year-old woman complains of a left posterior toothache for the past 2 weeks that she treated with saltwater gargles. However, over the past 24 h, she has had fever and difficulty opening her mouth while talking or swallowing. On examination, the patient has a fever of 101°F, with redness of the left submandibular region extending to the left side of her throat. She is sitting up but is anxious and drooling and has some inspiratory stridor. The physician states that the infection in the mouth has spread to the neck and may ultimately enter the chest.

► What is the most likely diagnosis?
► What is the anatomical mechanism for this condition?

ANSWERS TO CASE 52:

Dental Abscess/Ludwig Angina

Summary: A 45-year-old woman had a left molar toothache for 2 weeks but now has fever, trismus, and dysphagia. There is a left submandibular inflammation extending to the left side of the throat. She is sitting but is anxious and drooling and has some inspiratory stridor. This infection may track from the mouth to the neck to the chest.

- **Most likely diagnosis:** Submandibular cellulitis (Ludwig angina)

- **Anatomical mechanism for this condition:** A dental (molar) abscess that has tracked inferiorly from the submandibular space to impinge on the trachea

CLINICAL CORRELATION

Dental abscesses are relatively common occurrences and typically are self-limited or easily treated with antibiotics such as penicillin. Occasionally, an infection involving the molar teeth may extend into the submandibular space (Ludwig angina) and affect the trachea or carotid sheath contents. Fever, painful edema, limited neck mobility, drooling, and difficulty opening the mouth are clinical findings. The infection can also extend inferiorly into the mediastinum (mediastinitis). The inspiratory stridor in this case may indicate tracheal compression. In such cases, laryngoscopy may lead to laryngospasm and complete airway obstruction. Lateral neck radiographs or CT imaging are helpful in the diagnosis. The best treatment is intravenous antibiotics, airway protection (intubation if needed), and operative drainage of the abscess.

APPROACH TO:

The Oral Cavity

OBJECTIVES

1. Be able to list the layers of the deep cervical fascia

2. Be able to describe the structures in the floor of the mouth and submandibular space and its communications with the spaces of the neck

3. Be able to describe the route of spread of infection from the oral cavity into the thorax

DEFINITIONS

STRIDOR: A high-pitched whispering sound with respiration that indicates obstruction of the airway

TRISMUS: Sustained contraction of the masseter muscle, leading to "lockjaw"

DYSPHAGIA: Difficulty or pain with swallowing

LIGAMENTUM NUCHAE: A thickened extension of the supraspinal ligament into the neck.

DISCUSSION

The **deep cervical fascia** consists of connective tissue sheets that enclose and support various structures in the neck. Deep to the superficial fascia and **platysma**, the investing fascia (the superficial layer of deep fascia) **encircles the neck** and splits to enclose the **SCM** and the **trapezius muscles** and attaches to the **ligamentum nuchae** posteriorly. Superiorly, it attaches to the **hyoid bone, mandible, and base of the skull;** inferiorly, it attaches to the **acromion, clavicle, and manubrium of the sternum.** The **prevertebral fascia** surrounds the cervical vertebral column, the spinal cord, and the pre- and paravertebral musculatures. It attaches to the base of the skull superiorly and the ligamentum nuchae posteriorly, and blends with the anterior longitudinal ligament of the vertebral column in the thorax. The **pretracheal fascia** surrounds the larynx, trachea, esophagus, thyroid, and parathyroid glands and splits to enclose the **infrahyoid** (strap) **muscles** of the neck. It is attached superiorly to the hyoid bone and inferiorly blends with the fibrous pericardium in the thorax. Posteriorly and superiorly, it is continuous with the **buccopharyngeal fascia.** The carotid **sheath** is usually described as having originated in the investing, prevertebral, and pretracheal layers.

Between the prevertebral and buccopharyngeal fasciae lies the **retropharyngeal space** ("danger space"). This space is a pathway for spread of infection to the thorax, possibly resulting in cardiac tamponade. Within the pretracheal fascia is a potential space filled with loose areolar connective tissue called the **visceral space** (Figure 52-1).

The **submandibular space** lies between the mucosa of the floor of the mouth and the mylohyoid and hyoglossus muscles. The root of the tongue lies medially, and the inner surface of the mandible lies laterally. The space contains the **sublingual gland and ducts,** a portion of the submandibular gland and its duct, and the lingual and hypoglossal nerves. A cleft exists between the mylohyoid and hyoglossus muscles, through which the submandibular gland wraps around the posterior border of the mylohyoid muscle. The roots of the **posterior molar teeth** are close to the inner surface of the mandible, thus increasing the risk for dental abscesses **spreading into the submandibular space. Infectious material can thus spread inferiorly into the visceral space through the cleft between the mylohyoid and hyoglossus muscles.**

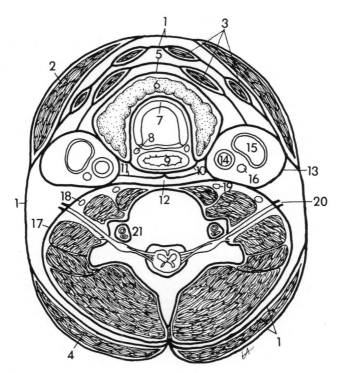

Figure 52-1. Compartments of the neck: 1 = investing fascia, 2 = sternocleidomastoid muscle, 3 = infrahyoid muscle, 4 = trapezius muscle, 5 = visceral (pretracheal) fascia, 6 = thyroid gland, 7 = trachea, 8 = recurrent laryngeal nerve, 9 = esophagus, 10 = buccopharyngeal fascia, 11 = alar fascia (present only in upper pharynx), 12 = retropharyngeal (retroesophageal) space, 13 = neurovascular (carotid) sheath, 14 = common carotid artery, 15 = internal jugular vein, 16 = vagus nerve, 17 = prevertebral fascia, 18 = phrenic nerve, 19 = sympathetic trunk, 20 = roots of the brachial plexus, 21 = vertebral artery. (*Reproduced, with permission, from the University of Texas Health Science Center, Houston Medical School.*)

COMPREHENSION QUESTIONS

52.1   A 67-year-old man developed a dental abscess that he ignored for 2 weeks. At that time, he developed severe chest pain due to infection of the mediastinum. Through which pathway did the infection most likely spread to the mediastinum?

A. Masticator space

B. Pretracheal space

C. Retropharyngeal space

D. Suprasternal space

52.2   A dentist uses local anesthesia to prepare for a procedure on a lower molar tooth. Which of the following nerves is the dentist blocking?

A.   Submental

B.   Maxillary

C.   Mandibular

D.   Vagus

52.3   A 24-year-old male was involved in a knife fight in a bar. He appeared in the emergency department with a 2-cm laceration in the anterolateral neck. The wound was superficial, but the physician observed muscle fibers just deep to the superficial fascia. Which of the following muscles was observed?

A.   Platysma

B.   Sternocleidomastoid

C.   Omohyoid

D.   Trapezius

E.   Thyrohyoid

ANSWERS

52.1   **C.** The major pathway between the infections of the neck and the chest is through the retropharyngeal space, which is a potential space between the prevertebral layer of fascia and the buccopharyngeal fascia surrounding the pharynx.

52.2   **C.** Dental anesthesia involving the lower molar teeth is called a **lower mandibular block**. The nerve affected is the inferior alveolar nerve branch of the mandibular nerve, which is a branch of V3.

52.3   **A.** The platysma muscle is a wide flat muscle that covers the anterolateral region of the neck.

ANATOMY PEARLS

▶ The submandibular space is continuous with the visceral space in the neck.

▶ The investing, pretracheal, and prevertebral deep cervical fasciae contribute to the carotid sheath.

▶ The major pathway for infection between the neck and the chest is the retropharyngeal space.

REFERENCES

Gilroy AM, MacPherson BR, Ross LM. *Atlas of Anatomy*, 2nd ed. New York, NY: Thieme Medical Publishers; 2012:530–531, 576, 606–609.

Moore KL, Dalley AF, Agur AMR. *Clinically Oriented Anatomy*, 7th ed. Baltimore, MD: Lippincott Williams & Wilkins; 2014:985–989.

Netter FH. *Atlas of Human Anatomy*, 6th ed. Philadelphia, PA: Saunders; 2014: plates 26, 67, 75.

A 3-day-old female is noted to have excessive tearing of the left eye and a small, firm, pea-size mass at the inferior region of the junction between the eye and the nose (oculonasal junction). The mass is not inflamed, and the infant is otherwise in good health and feeding well.

► What is the most likely diagnosis?
► What is the anatomical explanation for this disorder?

ANSWER TO CASE 53:
Lacrimal Sac Enlargement

Summary: A 3-day-old healthy infant has excessive tearing of the left eye and a small, firm, pea-size mass inferior to the medial canthus.

- **Most likely diagnosis:** Lacrimal sac enlargement (dacryocystocele)
- **Anatomical explanation of the disorder:** Congenital atresia of ducts draining into or out of the lacrimal sac

CLINICAL CORRELATION

The tear drainage system begins at the lacrimal puncta at the medial portion between the upper and lower eyelids. The puncta open into lacrimal canaliculi, which terminate at the lacrimal sac and, in turn, are drained by the nasolacrimal duct. The nasolacrimal duct develops from a solid cord of cells that recanalizes to establish the lumen of the duct and terminates in the inferior nasal meatus. Atresia of the duct (due to failure to recanalize) occurs in 1 to 3 percent of newborns. Atresia of the lacrimal canaliculi presents with excessive tearing and without a mass. Nasolacrimal duct atresia presents as a mass due to enlargement of the lacrimal sac, and the mass accompanied by excessive tearing suggests atresia of the canaliculi and the nasolacrimal duct. Massage of the nasolacrimal duct region with vigilant monitoring is the usual treatment, and most cases resolve by age 6 months. Persistent obstruction after age 9 months warrants intervention, such as nasolacrimal duct probing. Care must be exercised to avoid creating a false tract. Because the canaliculi and duct are obstructed in this case, duct probing is indicated.

APPROACH TO:
The Lacrimal Gland

OBJECTIVES

1. Be able to describe the anatomy of the lacrimal gland
2. Be able to describe the pathway for drainage of tears from the ocular globe to the nasal cavity

DEFINITIONS

LACRIMAL DUCT PROBING: Outpatient surgical procedure whereby a thin metal probe is used to cannulate the lacrimal duct, which is presumably occluded

DACRYOCYSTOCELE: Enlargement of the lacrimal sac

CANTHUS: Angle formed by the upper and lower eyelids

ATRESIA: Absence of a normal opening due to a developmental defect

DISCUSSION

The **lacrimal gland** is located in a shallow fossa at the superolateral aspect of the orbit (Figure 53-1). Approximately 12 small lacrimal ducts drain each gland, whose secretions or tears enter the **conjunctival sac superolaterally** at the superior conjunctival fornix and wash over the surface of the eye in an inferomedial direction, aided by the blinking action of the eyelids. The lacrimal gland is innervated by **autonomic nerves**, with the secretomotor fibers being a part of **CN VII**, whereas the sympathetic fibers are vasoconstrictive. Both types of fibers reach the gland through the lacrimal branch of the ophthalmic division of CN V.

Tears accumulate at the medial angle of the eye in the lacrimal lake. At the medial ends of the upper and lower eyelids, a small elevation, the **lacrimal papilla,** has an opening or **punctum** that leads to the **lacrimal canaliculus.** The two canaliculi terminate in the **lacrimal sac,** a blind-ended membranous structure continuous inferiorly with the **nasolacrimal duct.** The duct passes through the nasolacrimal canal of each maxilla and terminates in the **inferior nasal meatus,** the space bounded by the inferior nasal concha. Tears then pass to the nasopharynx and are swallowed.

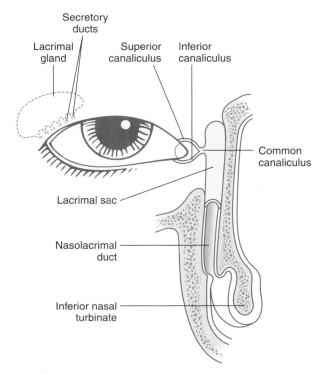

Figure 53-1. Lacrimal drainage system.

COMPREHENSION QUESTIONS

53.1 A 30-year-old woman underwent blunt trauma to the left eye and discovered that she was unable to secrete tears from that eye. Which of the following is the most likely location of the injury?

A. Medial superior orbit

B. Medial inferior orbit

C. Lateral superior orbit

D. Lateral inferior orbit

E. Adjacent to nasal bridge

53.2 A clinician places blue dye into the right eye to assess the patency of the tear duct system. Where should one look to see the eventual flow of the dye, assuming the lacrimal duct system is patent?

A. Superior nasal meatus

B. Middle nasal meatus

C. Inferior nasal meatus

D. Oral cavity

E. Subclavian vein

53.3 A 5-year-old boy is noted to have severe pain, swelling, and redness around his right eye. He has been diagnosed with periorbital cellulitis with probable spread of the infection to the brain. Which of the following routes best describes the probable avenue of spread to the brain?

A. Through the cribriform plate into the meningeal space

B. Facial vein to ophthalmic vein to cavernous sinus into the dural space

C. Frontal sinus into the sagittal sinus and into the subarachnoid space

D. Facial canal through the internal auditory meatus to the posterior cranial fossa

ANSWERS

53.1 **C.** The lacrimal gland, which produces tears, is located in the superior and lateral aspects of the orbit.

53.2 **C.** The tears flow through the puncta in the medial inferior aspect of the eyelid and travel through the lacrimal duct into the inferior nasal meatus.

53.3 **B.** Infections involving the periorbital space can penetrate through the ophthalmic vein into the cavernous sinus and into the dural space, leading to meningitis. Thus, prompt antibiotic therapy is critical with this infection.

ANATOMY PEARLS

▶ The lacrimal gland is located at the superolateral portion of the orbit.

▶ The parasympathetic secretomotor fibers originate in CN VII.

▶ Tears are produced by the lacrimal gland and drain through the lacrimal puncta into the lacrimal sac and through the nasolacrimal duct.

▶ The nasolacrimal duct terminates in the inferior nasal meatus.

▶ The most common cause of excessive tearing in a newborn is under-development of the lacrimal duct, which is usually treated with expectant management.

REFERENCES

Gilroy AM, MacPherson BR, Ross LM. *Atlas of Anatomy,* 2nd ed. New York: Thieme Medical Publishers; 2012:545.

Moore KL, Dalley AF, Agur AMR. *Clinically Oriented Anatomy,* 7th ed. Baltimore, MD: Lippincott Williams & Wilkins; 2014:892–893.

Netter FH. *Atlas of Human Anatomy,* 6th ed. Philadelphia, PA: Saunders; 2014: plates 83–84.

CASE 54

A 9-year-old boy is brought to the pediatrician because of severe headaches over the past week. The headaches were initially present only in the morning but over the last 48 h have become constant, lasting all day long. The child has also been vomiting and complains of problems "seeing." On examination, his temperature is 98°F and his heart rate is 80 beats/min. The child appears to be lethargic and is sensitive to the lights in the room. He has some neck rigidity, and his gait seems to be unsteady. Otherwise the examination is normal. A CT scan of the head is performed urgently and reveals bilateral enlargement of the frontal horns of the lateral ventricles and the third ventricle.

▶ What is the most likely diagnosis?
▶ Where is the obstruction most likely located?

ANSWER TO CASE 54:

Hydrocephalus

Summary: This is a 9-year-old child with progressive headache over 1 week, vomiting, and visual disturbances. On examination, the child has lethargy, photosensitivity, neck rigidity, and an unsteady gait. CT imaging shows lateral and third ventricle enlargement.

- **Most likely diagnosis:** Hydrocephalus

- **Obstruction:** Most likely the aqueduct of Sylvius leading to bilateral lateral ventricle and third ventricle enlargement.

CLINICAL CORRELATION

Hydrocephalus is the accumulation of cerebrospinal fluid in the brain. This increased CSF can be due to increased production, disturbance in flow, or altered absorption. Symptoms depend on the age of the patient and the rapidity of development. In this case, the child has an acute onset (less than 1 week) leading to more dramatic symptoms. If the hydrocephalus developed slowly such as over months, the only symptoms may be a vague headache and memory problems. With acute hydrocephalus, the most common complaint is headache. Children often are lethargic or have insomnia; if these symptoms are significant, they can also lead to vomiting. This child also has mild sensitivity to light (photophobia) and neck stiffness (rigidity), which are indications of meningeal irritation. Acute meningitis or blood in the CSF can also cause these symptoms. A brain tumor is another possibility; imaging of the brain is important to rule out the condition. In this case, significant hydrocephalus is identified on the CT scan. Because the lateral ventricles and the third ventricle are dilated, the most likely location of the obstruction is at the aqueduct of Sylvius. The treatment is to relieve the CSF accumulation, which usually involves a shunt placement procedure; a ventriculoperitoneal route is most commonly used. Surgical relief of the obstruction may also be attempted.

APPROACH TO:

Ventricular System of the Brain

OBJECTIVES

1. Be able to describe the location of the ventricles and circulation of CSF

2. Be able to identify the choroid plexus in the lateral, third and fourth ventricles, and the arachnoid granulations in the superior sagittal sinus.

DEFINITIONS

CHOROID PLEXUS: A structure in the ventricles of the brain where cerebrospinal fluid (CSF) is produced. The choroid plexus consists of modified ependymal cells.

ARACHNOID GRANULATIONS: Protrusions of arachnoid mater through the dura mater into the venous sinuses of the brain, allowing CSF to exit the subarachnoid space and return to the venous system. Most of the arachnoid granulations can be found in the superior sagittal sinuses.

DISCUSSION

There are four ventricles in the brain. The **lateral ventricles** are located within the cerebral hemispheres. The **third ventricle** is located between the diencephalons, and the **fourth ventricle** is located between the brainstem and the cerebellum (see Figure 54-1).

The paired lateral ventricles are the largest in the ventricle system of the brain and are each composed of one body and three horns: the anterior or frontal horn extends into the frontal lobe, the posterior or occipital horn extends into the occipital lobe, and the lateral or temporal horn extends into the temporal lobe. The body of the lateral ventricle is the central portion and is found in the parietal lobe just posterior to the frontal horn. The inner surface of the lateral ventricles is covered by a thin epithelial membrane called the **ependyma**. In the central portion and temporal horns of the ventricle, the ependyma is folded into the cavity with capillaries to form the **choroid plexus**, which produces CSF. CSF produced in the lateral ventricles flows into the third ventricle through interventricular foramen of Monro.

The third ventricle is a narrow, slitlike space. The choroid plexus is found in the superior part of this ventricle close to the interventricular foramen. The cerebral aqueduct of Sylvius conducts CSF from the third ventricle to the fourth ventricle.

The fourth ventricle has a tentlike configuration. The rhomboid fossa of the brainstem forms the floor, and the superior and inferior medullary velum of the cerebellum form the roof of this ventricle. The choroid plexus is found in the inferior part of this ventricle. From the fourth ventricle, CSF flows into the subarachnoid space through the foramina of Magendie and Luschka.

After the CSF circulates into the subarachnoid space and surrounds the brain and spinal cord, it is drained into the venous sinuses via the arachnoid granulations located mostly in the superior sagittal sinus (see Figure 54-2).

Increasing production of CSF, decreasing the absorption of CSF, or blocking the flow of CSF will cause it to accumulate in the ventricles, leading to hydrocephalus.

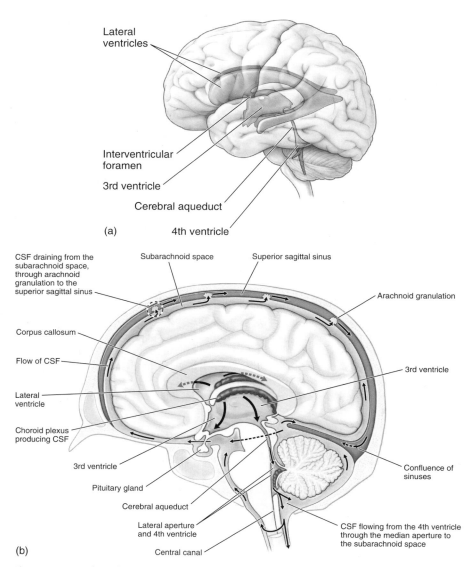

Figure 54-1. (a) Three-dimensional lateral view of the ventricles of the brain; (b) location and circulation of the cerebrospinal fluid (CSF). (*Reproduced, with permission, from Morton DA, Foreman KB, Albertine KH. The Big Picture: Gross Anatomy. New York: McGraw-Hill Medical, 2011. Figure 16-2.*)

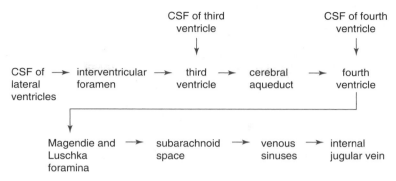

Figure 54-2. Flowchart showing circulation of CSF in the subarachnoid space.

COMPREHENSION QUESTIONS

54.1 A 32-year-old woman is seen in the emergency department complaining of severe headaches and difficulty walking. She is noted to be lethargic. On CT imaging of the head, she is noted to have enlargement of bilateral lateral ventricles and the third ventricle. The fourth ventricle is noted to be normal. The most likely obstruction in this patient is the

    A. Foramen of Luschka

    B. Foramen of Magendie

    C. Cerebral aqueduct

    D. Interventricular foramen of Monro

    E. Cisterna magna

54.2 Lumbar puncture may be performed to withdraw CSF for lab examination. During this procedure, which the following structures will the needle pass through before it enters the subarachnoid space?

    A. Supraspinous ligament

    B. Anterior longitudinal ligament

    C. Posterior longitudinal ligament

    D. Intertransverse ligament

    E. Pia mater

54.3 During the lumbar spinal puncture, the needle usually should be inserted in the midline between the spinal processes of the

    A. T12 and L1 vertebrae

    B. L1 and L2 vertebrae

    C. L2 and L3 vertebrae

    D. L3 and L4 vertebrae

    E. L5 and S1 vertebrae

ANSWERS

54.1 **C.** This patient likely has an obstruction of the cerebral aqueduct of Sylvius since this is the conduit between the third and fourth ventricles. Thus, there is dilation of both lateral ventricles and the third ventricle, but a normal fourth ventricle.

54.2 **A.** During the lumbar puncture, the needle will sequentially pass the supraspinous ligament, the interspinous ligament, the ligamentum flavum, the dura mater, and the arachnoid mater to enter the subarachnoid space. The anterior and posterior longitudinal ligament is located in the anterior and posterior surface of vertebral body and deep to the vertebral canal and should not be reached; the pia mater is a thin layer of meninges on the surface of the spinal cord. During lumbar puncture, the needle is inserted below the inferior end of the spinal cord and should therefore not reach the pia mater.

54.3 **D.** In an adult, the inferior end of the spinal cord is at the level of the inferior border of the L1 vertebra. Therefore, to protect the spinal cord, it is safest to insert the needle between the L3 and L4 or the L4 and L5 vertebrae.

ANATOMY PEARLS

▶ There are three layers of meninges that surround the spinal cord and brain, and CSF fills the subarachnoid space, which is located between the arachnoid and pia maters.

▶ CSF is produced by the choroid plexus found in the four ventricles of the brain and flows back to the venous system through the arachnoid granulations. CSF surrounds the central nervous system to protect the brain and spinal cord.

▶ Laboratory examination of a sample of CSF can help us diagnose bacterial infections, viral infections, and tumors in the central nervous system.

▶ There are no lymphatic vessels in the central nervous system, and the CSF acts as the lymph of the brain and spinal cord.

REFERENCES

Gilroy AM, MacPherson BR, Ross LM. *Atlas of Anatomy*, 2nd ed. New York, NY: Thieme Medical Publishers; 2012:628–630, 632–633.

Moore KL, Dalley AF, Agur AMR. *Clinically Oriented Anatomy*, 7th ed. Baltimore, MD: Lippincott Williams & Wilkins; 2014:878–881, 886–887.

Netter FH. *Atlas of Human Anatomy*, 6th ed. Philadelphia, PA: Saunders; 2014: plates 109–110.

A 19-year-old star college football player was tackled from the right side during a game, in which the opposing player hit the patient's legs with the shoulder. The patient said that he immediately heard a "popping" sound. The player started to complain of stiffness and swelling in his knee, which he felt was "giving out." He was carted off the field by emergency medical services (EMS), and on examination, he had an unstable gait and lost full range of motion in his right knee. His knee was tender to palpation. He has no past medical history and is otherwise healthy.

► What is the most likely diagnosis?
► What knee structures are most likely injured?
► What skeletal and muscular components are involved in the injury?

ANSWER TO CASE 55:
Knee Injury

Summary: A 19-year-old male with no past medical history is tackled during a football game and is complaining of pain in his right knee after hearing a "popping sound." His gait is unstable and he has instability in his knee with twisting motion. In addition, he has lost his full range of motion.

- **Most likely diagnosis:** Injury to unhappy triad ("O'Donoghue's triad", "terrible triad").

- **Knee structures likely injured:** Anterior cruciate ligament, medial collateral ligament, and medial meniscus.

- **Skeletal components involved:** Patella, femur, and tibia. There are no muscles directly involved with this type of injury.

CLINICAL CORRELATION

The unhappy triad consists of damage to the anterior cruciate ligament (ACL), medial collateral ligament (MCL), and medial meniscus and is common among sports injuries when a player is tackled with great force such as in football, rugby, or soccer.

This athlete was playing in a football game and was otherwise healthy. He was clipped perpendicularly from the lateral aspect of the knee such that the force was directed from lateral to medial, thereby straining the MCL and medial meniscus. Along with this, the sudden inward twisting of the knee with the foot planted causes a strain on the ACL. Hearing a popping sound is a very common symptom with injury to the ACL. The patient's unstable gait and instability in twisting motion from side to side further confirm the suspicion that he damaged ligaments in the unhappy triad. Other common symptoms are pain and swelling in the knee immediately after injury.

APPROACH TO:
The Knee Joint

OBJECTIVES

1. Be able to describe the anatomy of the knee joint, including the articular surface, joint capsule, ligaments, and menisci

2. Be able to describe the various mechanisms of injury to the knee joint

DEFINITIONS

MENISCUS: A crescentlike fibrocartilaginous structure located in the knee joint between the condyles of the tibia and femur bones. The medial meniscus (C-shaped) and lateral meniscus (O-shaped) act as cartilaginous pads to absorb shock, fit the articular surfaces better, and increase the flexibility of the knee joint.

LATERAL MENISCUS: A smaller structure, separated from the lateral collateral ligament by the tendon of the popliteus muscle. It is more freely mobile than the medial meniscus.

DISCUSSION

The knee joint is the largest and most complex synovial joint in the body. It is formed by the femur bone, tibia bone, and patella. The fibula bone is not a part of the knee joint (Figure 55-1).

Articular Surfaces

The lateral and medial femoral condyles join with the lateral and medial tibial condyles to form the femorotibial articulations. The patella surface of the femur bone joins with the largest sesamoid bone the patella to form the femoropatellar articulation. The articular surfaces do not fit perfectly with each other.

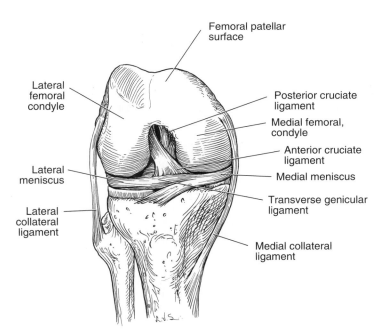

Figure 55-1. Anterior view of the right knee. (*Reproduced, with permission, from Lindner HH. Clinical Anatomy. East Norwalk, CT: Appleton & Lange, 1989:615.*)

Articular Capsule and Cavity

The **fibrous capsule** is a thin but strong membrane that attaches to the articular margins of the condyles of the femur and tibia, and the fibrous capsule is fused with the patellar ligament anteriorly. The **synovial capsule** lines the inner surface of the fibrous capsule, and it reflects on the interior between the tibia and femur bones to form the synovial folds, like the infrapatellar fold, alar fold, and similar structures. Between the muscles and the tendons that surround the knee, the synovial capsule projects exteriorly through a break in the fibrous capsule to form the synovial bursae, such as the suprapatella and infrapatellar bursae. Bursae are an extension of the articular cavity and can reduce friction when the muscle or tendon moves on the surface of the knee.

The **articular cavity** is the space between the articular surfaces and the articular capsule. In the knee joint, the articular cavity is relatively narrow, which contributes to the joint's stability.

Ligaments

There are two groups of ligaments to strengthen the knee joint: the extracapsular ligaments and intraarticular ligaments (see Figure 55-2):

1. Extra capsular ligaments (external ligaments) are located on the external surface of the joint capsule:

   A. **Patellar ligament:** the inferior part (below the patella) of the quadriceps tendon. It is the strongest ligament of the knee joint and protects the joint from the anterior aspect.

   B. **Lateral collateral ligament:** a cordlike, strong ligament that extends from the lateral epicondyle of the femur to the lateral surface of the fibula and is separated from the lateral meniscus by the tendon of popliteus muscle. The lateral collateral ligament prevents adduction of the knee joint.

   C. **Medial collateral ligament:** a flat, bandlike ligament that extends from the medial epicondyle of the femur to the medial surface of the tibia. Deep fibers

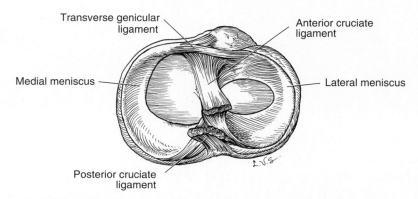

Figure 55-2. The fully flexed right knee, showing ligaments of the knee joint. (*Reproduced, with permission, from Lindner HH. Clinical Anatomy. East Norwalk, CT: Appleton & Lange, 1989:615.*)

of this ligament firmly attach to the medial meniscus; thus, the ligament is often damaged together with the medial meniscus. Both of the collateral ligaments strengthen the knee joint on the sides. The medial collateral ligament prevents abduction of the knee joint.

D. **Oblique popliteal ligament:** the posterior ligament of the knee that extends superiorly and laterally from the medial epicondyle of the tibia to the posterior surface of the joint capsule. It strengthens the capsule of knee joint posteriorly.

E. **Arcuate popliteal ligament:** a small ligament that extends from the posterior surface of the fibular head and crosses the tendon of the popliteus to the posterior surface of the capsule. Just like the oblique popliteal ligament, it strengthens the capsule of knee joint posteriorly.

2. Intraarticular ligaments (internal ligaments)

A. **Cruciate ligaments:** two ligaments located in the middle of the joint that cross each other forming what looks like the letter "X". The anterior cruciate ligament connects the tibia (anterior to the intercondylar eminence) to the medial surface of the lateral condyle of the femur. The posterior cruciate ligament connects the tibia (posterior to the intercondylar eminence) to the lateral surface of the medial condyle of the femur. The cruciate ligaments prevent the tibia from sliding anteriorly and posteriorly on the femur, providing stability to the knee joint. In addition to the cruciate ligaments, there are other small ligaments in the knee joint, such as the transverse ligament, which are functionally less important to the joint.

Movement of the Knee Joint

Two **menisci are located** between the femur and the tibia and contribute to the flexibility of the knee joint (see Figure 55-3).

The knee joint is a hinge-type joint that permits flexion and extension. When the joint is flexed, the round posterior surface of the medial and lateral condyles of the femur contact the condyles of the tibia and permit slight medial and lateral rotation.

The knee joint is mainly supplied by five genicular arteries that originate from the popliteal artery. The superior medial genicular, superior lateral genicular, inferior medial genicular, and inferior lateral genicular arteries form the genicular anastomosis around the knee (between the muscles and the bones), and the middle genicular artery crosses the posterior capsule to supply the interior structures of the knee.

During sports and other high-contact activities, the knee joint can often be subjected to abnormal forces from the anterior and lateral directions, causing tears of the anterior cruciate ligament. Forced abduction of the knee joint can tear the medial collateral ligament, and because the medial collateral ligament is firmly attached to the medial meniscus (especially with the limited movement between the femur and tibia bones), the medial meniscus is commonly injured with the medial collateral ligament. The anterior cruciate ligament, medial collateral ligament, and medial meniscus are frequently damaged simultaneously; this event is commonly known as the "unhappy triad of the knee."

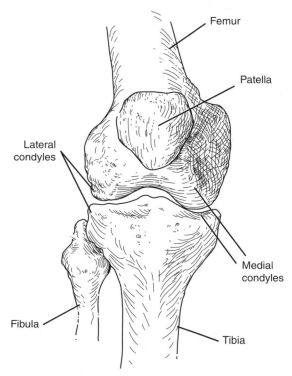

Figure 55-3. Superior aspect of the right tibia showing menisci of the knee joint. [*Reproduced, with permission, from Lindner HH. Clinical Anatomy. East Norwalk, CT: Appleton & Lange, 1989:613 (Figure 49-1).*]

COMPREHENSION QUESTIONS

55.1 Which of following structures of the knee joint contribute to its mobility?

   A. Lateral collateral ligament

   B. Medial collateral ligament

   C. Patellar ligament

   D. Anterior cruciate ligament

   E. Meniscus

55.2 The medial meniscus is firmly attached to which ligament?

   A. Fibular collateral

   B. Tibial collateral

   C. Anterior cruciate

   D. Posterior meniscofemoral

   E. Patellar

55.3 The knee joint
   A. Is "unlocked" by the plantaris
   B. Contains an anterior cruciate ligament to prevent hyperextension
   C. Has a lateral meniscus attached to the fibular collateral ligament
   D. Is flexed by the quadraceps femoris
   E. Is extended by the hamstring muscles

ANSWERS

55.1 **E.** Menisci are located between the femur and tibia bones and allow the artic-ular surfaces to fit each other better. They also separate the joints into two parts: femoral-meniscus and meniscus-tibial joints, but the meniscus does not contribute to the mobility of the joint.

55.2 **B.** The tibial collateral ligament is a broad, strong ligament that extends between the medial epicondyle of the femur and the medial surface of the upper tibia bone. The medial meniscus is attached to this ligament.

55.3 **B.** The function of the ACL is to prevent hyperextension in the knee joint. The "unlocking" muscle for the knee joint is the popliteus muscle, which separates the lateral meniscus from the fibular collateral ligament. The quad-riceps femoris is the extensor of the knee joint, and the hamstring muscles are the flexor of the knee joint.

ANATOMY PEARLS

▶ The knee joint is the largest and most complex joint in the human body.

▶ The knee joint includes three articulations between the femur, tibia, and patella.

▶ The extracapsular ligaments and intraarticular ligaments stabilize the knee joint, and the menisci contribute to the flexibility of the knee joint.

▶ The anterior cruciate ligament (ACL), medial meniscus, and medial col-lateral ligament are commonly injured during sports.

REFERENCES

Gilroy AM, MacPherson BR, Ross LM. *Atlas of Anatomy*, 2nd ed. New York, NY: Thieme Medical Publishers; 2012:408–414.

Moore KL, Dalley AF, Agur AMR. *Clinically Oriented Anatomy*, 7th ed. Baltimore, MD: Lippincott Williams & Wilkins; 2014:634–643, 662–665.

Netter FH. *Atlas of Human Anatomy*, 6th ed. Philadelphia, PA: Saunders; 2014: plates 494–498.

CASE 56

An 18-year-old man is seen in the emergency department complaining of a "stomachache all over" and fever. He reports that 2 days ago, he had some soreness around the umbilicus, and then yesterday, the pain seemed to go to the right lower abdomen. Today, he complains of pain throughout the abdomen with fever and chills. He is not hungry. On examination, his temperature is 102°F, heart rate is 110 beats/min, and blood pressure is 130/90 mmHg. An abdominal examination reveals a distended abdomen. There are hypoactive bowel sounds on auscultation. The patient has generalized tenderness throughout the abdomen with involuntary guarding and rebound tenderness.

► What is the most likely diagnosis?
► What is the explanation for the change in location of the pain?
► What is the mechanism for the rebound tenderness?

ANSWER TO CASE 56:
Peritoneal Irritation

Summary: An 18-year-old man is seen for a progression of abdominal pain that began periumbilically, and then spread to the right lower quadrant, followed by generalized abdominal pain. He has fever, hypoactive bowel sounds, and involuntary guarding with rebound tenderness.

- **Most likely diagnosis:** Acute appendicitis, likely ruptured with generalized peritonitis.

- **Explanation for change in location of pain:** Originally, the pain from the appendicitis is referred to the umbilicus (visceral sensation), and then as the appendicitis becomes more acute and inflamed, the parietal peritoneum is affected and localizes to the right lower quadrant. Finally, the perforation leads to purulent material throughout the abdominal cavity with peritoneal irritation causing rebound tenderness.

- **Mechanism for rebound tenderness:** Quick release of pressure from the clinician's hand leads to "rebounding" of the peritoneum, which, if inflamed, will cause pain.

CLINICAL CORRELATION

This young man has the typical presentation of acute appendicitis that has progressed initially from engorgement (visceral pain) to inflammation affecting the parietal peritoneum (somatic pain), and finally to frank rupture of the appendix. Pus is released into the entire peritoneal cavity, leading to generalized pain and rebound tenderness. Visceral pain is typically within the walls of hollow organs and stimulated by stretching, distension, or contractions. It is poorly localized and usually felt in the midline. In this case the distension of the appendix leads to a poorly defined periumbilical pain. Further questioning may lead to a description of deep, dull aching or cramping. When the appendix becomes inflamed and the inflammation on its surface touches the parietal peritoneum, there is more localized pain. This pain is described as sharper, aggravated by stimulation of the parietal peritoneum such as movement, coughing, or walking. In eliciting rebound tenderness, the physician presses deep on the abdomen and then quickly removes the hand (or pressure), and the patient experiences a sudden onset of pain on release of the pressure, rather than from the pressure itself. This is due to peritoneal irritation, and the pain occurs because the peritoneum rebounds back, activating sensory fibers, when the pressure is suddenly released. Other indications of peritoneal irritation include pain on percussion of the abdomen.

APPROACH TO:
Peritoneum

OBJECTIVES

1. Be able to define the differences between visceral peritoneum, parietal peritoneum, and a mesentery (peritoneum ligament or omentum)

2. Be able to define the peritoneal cavity, greater sac, lesser sac, and their contents (if any)

3. Be able to describe the differences in the sensory innervation of the visceral versus parietal peritoneum

DEFINITION

REFERRED PAIN: Perception of pain superficially that is arising from a deeper, often distant source

DISCUSSION

The **peritoneum** consists of a thin serous membrane composed of a simple squamous epithelium called **mesothelium**, and a thin layer of loose connective tissue, rich in elastic fibers. The peritoneum is divided into a portion that lines the inferior surface of the diaphragm and the abdominal and pelvic walls, the **parietal peritoneum**, and the portion that covers all or a part of the abdominopelvic viscera, the **visceral peritoneum** (see Figure 56-1).

Another peritoneal structure is a double-peritoneal sheet with a connective core, called a **mesentery**. The connective tissue core may contain a large amount of fat, serving the body as a major storage site for fat. Blood vessels and nerves passing to and from the viscera and the posterior body region are also located within the connective tissue core. These double-peritoneal sheets are sometimes termed **ligament** or **omentum**.

The space between the parietal and visceral peritoneum is called the **peritoneal cavity**. The peritoneum produces a small amount of serous fluid called **peritoneal fluid**, which lubricates movement of the viscera suspended in the peritoneal cavity. The peritoneal cavity in subdivided into the large **greater sac** extending from the diaphragm superiorly, to the pelvic cavity inferiorly. A smaller **lesser sac** or **omental bursa** is found posterior to the liver and stomach. It communicates with the greater sac via the **omental foramen** (epiploic foramen of Winslow). The peritoneal cavity of the male is closed, but that of the female is open to the outside via the uterine tubes, uterus, and vagina.

The sensory innervation of the peritoneum is important clinically. The parietal peritoneum of the central underside of the diaphragm (derived from the septum transversum) receives its sensory innervation from the **phrenic nerve (C3–C5)**. Innervation of the peritoneum on the underside of the diaphragm's periphery is provided by **spinal nerves T6 through T12**. Innervation of the peritoneum lining the

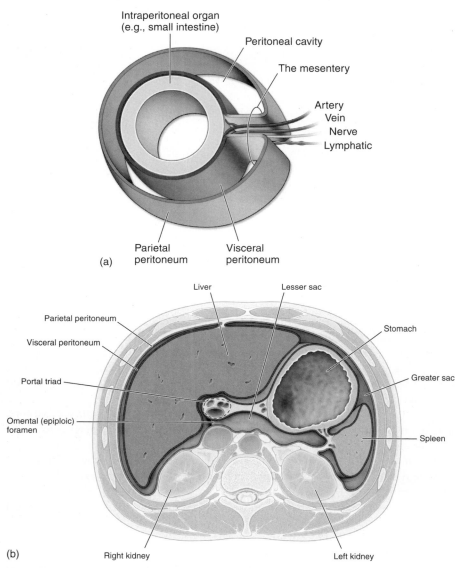

Figure 56-1. (a) Relationship between the mesentery and the peritoneal organs; (b) axial cross section of the peritoneum and mesentery. [*Reproduced, with permission, from Morton DA, Foreman KB, Albertine KH. The Big Picture: Gross Anatomy. New York: McGraw-Hill, 2011:99 (Figure 8-1B and C).*]

abdominal wall is provided by **spinal nerve T6 through T12 and L1**, while the peritoneum lining the pelvic wall is innervated by the **obturator nerve (L2–L4)**. These somatic nerves providing sensory innervation to the parietal peritoneum are essentially sensitive to pain, touch, temperature, and pressure. This latter sensation is the basis of **rebound tenderness** from an already irritated peritoneum. These somatic nerves from the parietal peritoneum provide an intense, well-localized sensation. Sensory innervation from the visceral peritoneum covering most of the abdomino-pelvic organs, as well as their mesenteries, are not sensitive to touch, temperature, or pressure, but are sensitive to ischemia, stretching, or tearing, such as from a swollen or distended organ. These visceral afferent nerves are described as being part of the autonomic nervous system (ANS), and travel back to the spinal cord via the sympathetic portion of the ANS. They convey a dull, poorly localized sensation.

Referred pain means the sensation of pain at a site different from its original source. Pain sensation originating from a gastrointestinal organ is often perceived at or near the midline. This is attributed to the fact that these organs are midline in origin. The clinically important referred pain involves both the visceral and somatic sensory nerves. For example, the visceral afferent fibers from the stomach travel to the spinal cord via the greater splanchnic nerves to reach the T5 through T9 levels of the spinal cord. Pain from the stomach is often perceived initially and somewhat vaguely at the epigastric midline, which, in turn, is supplied by spinal nerves T5 through T9. Visceral afferent fibers from the appendix enter the spinal cord at approximately the T10 level, and pain from a distended appendix is initially perceived at the periumbilical region which is typically supplied by the T10 spinal nerve. If the organ is inflamed and becomes distended, as is often the case, the adjacent parietal peritoneum may also became irritated. In such instances, the initially vague periumbilical discomfort can shift to a well-localized, intense right lower quadrant pain from the appendix itself. This well-localized pain may be accompanied by **muscular rigidity** or "guarding," which is a body reflex, while attempting to reduce peritoneal movement, which, in turn, may produce pain.

The mechanism for referred pain is not fully understood. It may be more complex than entry of sensory nerve fibers into the central nervous system (CNS) at a common spinal cord level (e.g., T10 for the periumbilical region and the appendix itself). For example, a common pathway that courses superiorly to the brain from the spinal cord may also be involved in the conscious perception of pain.

COMPREHENSION QUESTIONS

56.1 During a dissection, you inform your colleague that the surface of the peritoneum covering the stomach is composed of

    A. Simple squamous epithelium

    B. Simple cuboidal epithelium

    C. Simple columnar epithelium

    D. Stratified squamous epithelium

    E. Transitional epithelium

56.2 You point to the colon with your probe, and ask your colleague "What tissue am I touching?"

A. Visceral **pleura**

B. Parietal pleura

C. Mesentery

D. Parietal peritoneum

E. Visceral peritoneum

56.3 You ask your dissection colleague "What spinal nerves innervate the peritoneum on the underside of the central region of the diaphragm?"

A. C3 through C5

B. T5 through C9

C. T6 through T12

D. T10

E. L2 through L4

56.4 Continuing your discussion with your dissection colleague, you ask "Which spinal nerve innervates the umbilical region?"

A. T4 through T6

B. T8

C. T10

D. T12

E. L1

ANSWERS

56.1 **A.** The peritoneal epithelium is the simple squamous type.

56.2 **E.** The abdominal organs are covered in whole or in part by visceral peritoneum.

56.3 **A.** The sensory impulses of the central underside of the diaphragm are sent to segments C3 through C5 of the spinal cord, from which the phrenic nerve arises.

56.4 **C.** The dermatome at the level of the umbilicus is T10, and its sensory fibers are a part of the T10 spinal nerve.

ANATOMY PEARLS

► The epithelium of the peritoneum is simple squamous, called **mesothelium.**

► Parietal peritoneum lines the underside of the diaphragm and abdomino-pelvic walls and, if irritated, is characterized by intense, well-localized pain.

► Visceral peritoneum covers some, if not the entire, surface of abdomino-pelvic organs and is characterized by vague, dull pain if irritated.

► The sensory fibers of the appendix and umbilicus reach the T10 level of the spinal cord.

► Muscle rigidity or guarding helps reduce the pain from the parietal peritoneum by reducing movement.

REFERENCES

Gilroy AM, MacPherson BR, Ross LM. *Atlas of Anatomy,* 2nd ed. New York, NY: Thieme Medical Publishers; 2012:150–155.

Moore KL, Dalley AF, Agu AMR. *Clinically Oriented Anatomy,* 7th ed. Baltimore, MD: Lippincott Williams & Wilkins; 2014:159, 217–218.

Netter FH. *Atlas of Human Anatomy,* 6th ed. Philadelphia, PA: Saunders; 2014: plates 263–267.

A 64-year-old man complains of pain and weakness in his right shoulder for the previous 2 months. He states that the pain is worse when he tries to lift his arm and he has difficulty keeping his arm elevated for more than a few seconds. He denies any falls or trauma to the arm or shoulder. His right hand is dominant. On examination, he has minor pain on palpation of the right shoulder. The pain is increased with abduction of his arm greater than 90°. Additionally, he is unable to hold his arm in an abducted position and has weakness with external rotation. Following injection of lidocaine in the joint, his pain disappears, but the weakness continues.

▶ What is the most likely diagnosis?
▶ What anatomical structure is involved?

ANSWER TO CASE 57:

Rotator Cuff Tear

Summary: A 64-year-old man with no previous trauma is complaining of chronic pain and weakness in his dominant arm. He has pain with abduction in addition to weakness with external rotation on examination. Injection of a local anesthetic relieves his pain but not help the weakness that he has been experiencing.

- **Most likely diagnosis:** Rotator cuff tear

- **Anatomical structure likely involved:** The rotator cuff, which consists of the supraspinatus, infraspinatus, teres minor, and subscapularis muscles

CLINICAL CORRELATION

This elderly man has the typical findings of a rotator cuff tear. Although some patients may be asymptomatic, common complaints include pain and weakness with abduction. Rotator cuff injuries are very common, especially in patients over the age of 40. The rotator cuff may be torn acutely, such as with trauma, or it may be a chronic issue, with both degeneration secondary to age and repetitive stress contributing. The rotator cuff stabilizes the glenohumeral joint and facilitates various arm movements. The supraspinatus contributes to abduction of the arm, especially early abduction. The infraspinatus and teres minor are external rotators. The subscapularis rotates the arm internally. Lidocaine injection is helpful for diagnosis as it distinguishes rotator cuff tendinopathy from a tear. Lidocaine relieves pain in both injuries, but will improve strength in only tendinopathy.

APPROACH TO:

Scapulohumeral Muscles (Intrinsic Shoulder Muscles)

OBJECTIVES

1. Be able to describe the arrangement of intrinsic shoulder muscles

2. Be able to describe the actions of the rotator cuff muscles

3. Be able to understand the features of the shoulder joint

DISCUSSION

There are two groups of muscles surrounding the glenohumeral (shoulder) joint: **axioappendicular muscles** (extrinsic shoulder muscles) and **scapulohumeral muscles** (intrinsic shoulder muscles). The **extrinsic** muscles (total nine muscles) connect the upper limb to the trunk; the six **intrinsic** shoulder muscles (deltoid, teres major, supraspinatus, infraspinatus, teres minor, and subscapularis) originate from the pectoral girdle (scapula and clavicle), insert to the

Table 57-1 • ROTATOR CUFF MUSCLES

Muscle	Origin	Insertion	Nerve Innervation	Main Action
Supraspinatus	Supraspinal fossa of scapula	Superior facet of greater tubercle of humerus	Suprascapular	Initiates abduction of arm
Infraspinatus	Infraspinal fossa of scapula	Middle facet of greater tubercle of humerus	Suprascapular	Laterally rotate arm
Teres minor	Lateral border of scapula	Inferior facet of greater tubercle of humerus	Axillary	Laterally rotate arm
Subscapularis	Subscapular fossa of scapula	Lesser tubercle of humerus	Upper and lower subscapular	Medially rotate arm

humerus, and act on the shoulder joint. Four of the intrinsic shoulder muscles (supraspinatus, infraspinatus, teres minor, and subscapularis) are referred to as **rotator cuff muscles** (see Table 57-1) because their muscle fibers and tendons surround the capsule of the shoulder joint to form the musculotendinous rotator cuff (see Figure 57-1).

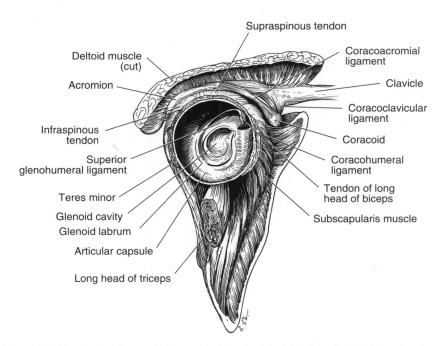

Figure 57-1. Muscles, tendons, and ligaments of the shoulder joint (lateral view). [*Reproduced, with permission, from Lindner HH. Clinical Anatomy. East Norwalk, CT: Appleton & Lange, 1989:528 (Figure 41-3).*]

The shoulder joint is a ball-and-socket synovial joint consisting of the large head of the humerus and small glenoid fossa of the scapula that fosters the mobility of the joint but also makes it relatively unstable.

The tendon of the supraspinatus crosses over the capsule superiorly, the tendons of the infraspinatus and the teres minor cross over the capsule posteriorly, and the tendon of the subscapularis crosses over the capsule anteriorly; these tendons (SITS) reinforce the joint capsule from three directions to protect the joint and give it stability.

Between the tendons of the rotator cuff muscles and the joint capsule are the **bursae**, which contain synovial fluid to reduce friction during muscle contractions. Most of the bursae directly communicate with the cavity of the shoulder joint.

Lesions or degeneration of the rotator cuff and related bursae are common causes of pain in the shoulder area.

COMPREHENSION QUESTIONS

57.1 A 44-year-old swimmer is seen in the physician's office for right shoulder pain for 1 year. A radiograph of the shoulder indicates that the supraspinatus tendon is calcified. What movement is most likely to cause discomfort in this patient?

    A. Medial rotation of the humerus

    B. Lateral rotation of the humerus

    C. Adduction of the humerus

    D. Abduction of the humerus

57.2 A 19-year-old man is involved in a motor vehicle accident and suffers a right proximal humeral fracture. The patient is noted to have numbness of the right lateral upper arm and also inability to abduct his arm. Which of following muscles is likely affected by this nerve injury?

    A. Teres minor

    B. Supraspinatus

    C. Subscapularis

    D. Teres major

    E. Infraspinatus

57.3 An important function of the rotator cuff is

    A. Depression of the clavicle

    B. Elevation of the clavicle

    C. Costoclavicular support

    D. Stabilization of the humeral head

    E. Protraction of the acromial angle

ANSWERS

57.1 **D.** This patient likely has rotator cuff syndrome due to repetitive wear-and-tear of the rotator cuff muscles. The rotator cuff consists of the supraspinatus, infraspinatus, teres minor, and subcapsularis muscles. The rotator cuff muscles act on abduction of the humerus. The other motions do not involve the rotator cuff muscles.

57.2 **A.** This patient likely has axillary nerve injury from blunt trauma to the quadrangular space. The **quadrangular space** is where the axillary nerve travels from anterior to posterior. It is bounded by the subscapularis muscle and teres minor, teres major, surgical neck of the humerus, and the long head of the triceps muscle. Injury to the axillary nerve affects the teres minor and deltoid muscles. The suprascapular nerve innervates the supraspinatus and infraspinatus muscles, and the subscapular nerve innervates the teres major and infraspinatus muscles.

57.3 **D.** Besides rotatory movement of the humerus in specific directions, the rotator cuff is an important stabilized of the shoulder.

ANATOMY PEARLS

▶ The rotator cuff is the musculotendinous structure that surrounds the shoulder joint. It is composed of four muscles and their tendons (SITS) that cross over the shoulder joint from three directions and contributes to the stabilization of the joint.

▶ Rotator cuff injury and degeneration in older people are common causes of shoulder pain.

REFERENCES

Gilroy AM, MacPherson BR, Ross LM. *Atlas of Anatomy*, 2nd ed. New York, NY: Thieme Medical Publishers; 2012:286–287, 290–297.

Moore KL, Dalley AF, Agur AMR. *Clinically Oriented Anatomy*, 7th ed. Baltimore, MD: Lippincott Williams & Wilkins; 2014:706–707, 712.

Netter FH. *Atlas of Human Anatomy*, 6th ed. Philadelphia, PA: Saunders; 2014: plates 405–408, 411, 413, 418.

A 48-year-old handyman complains of difficulty working due to increasing numbness, weakness, and pain in his right arm and hand. The paresthesias and weakness worsen when he lifts his arm over his head to perform tasks such as painting or hammering. The numbness also sometimes wakes him up at night and is worst on the volar side of his fourth and fifth fingers. He has also noticed that his right hand and fingers sometimes seem paler and colder than his left hand and fingers. He denies any history of trauma to the shoulder or arm and of any medical problems. On exam, some muscle wasting is evident in his right hand. Phalen's and Tinel's signs are negative. He is unable to complete the **elevated arm stress test** (EAST) or Roos test (the musculoskeletal shoulder flexibility or "hands up" test) because of heaviness and fatigue in his right arm. (In this test, the patient opens and closes his hands for 3 min while his arms are externally rotated and abducted to 90° with elbows flexed at 90°.)

▶ What is the most likely diagnosis?
▶ What anatomical structures are most likely affected?
▶ What are the common causes?

ANSWER TO CASE 58:

Thoracic Outlet Syndrome

Summary: A 48-year-old man has pain, paresthesias, and weakness in his right arm and hand worsened with arm abduction and at night. He also has signs of poor circulation in his right hand (pallor and coolness). On exam, EAST is positive, and muscle wasting is present in the right hand.

- **Most likely diagnosis:** Thoracic outlet syndrome

- **Anatomical structures likely affected:** Neural (brachial plexus branches), arterial (subclavian artery), and venous (subclavian vein) structures

- Common causes: Often associated with a cervical rib, but can also be caused by anomalous ligaments, hypertrophy of the anterior scalene muscle, or cervical trauma

CLINICAL CORRELATION

This man complains of neurological symptoms (numbness and tingling) as well as signs of arterial insufficiency (pallor, coolness) in his right arm and hand, suggesting compromise, or in this case compression, of neural and arterial structures–branches of the brachial plexus and the subclavian artery. (The subclavian vein can also be involved, producing venous signs such as swelling and edema.) The signs and symptoms are worsened with use, when more demand is placed on these structures; or with position, when the structures are further compressed. These structures run through the thoracic outlet, between the clavicle and the first rib. The brachial plexus and subclavian artery also run between the anterior and middle scalene muscles. (The subclavian vein is anterior to the anterior scalene, which is why it is less commonly involved.) The EAST as well as Adson's test and the costoclavicular maneuver can be helpful in detecting thoracic outlet syndrome but are not completely diagnostic. X-rays, MRI, and EMG can also be useful in demonstrating compression. Depending on the severity, treatment can range from stretching and physical therapy to surgery.

APPROACH TO:

Thoracic Outlet

OBJECTIVES

1. Be able to describe the anatomy of the thoracic outlet and the structures that exit through this opening

2. Be able to describe those adjacent structure(s) at risk from any pathologic process or structure at or near the thoracic outlet

DEFINITION

CERVICAL RIB: An abnormal extra rib (often bilateral) articulating with the C7 vertebra, and stretching structures exiting the thoracic outlet or that are nearby

DISCUSSION

While the superior opening of the thoracic cage is often called the **thoracic outlet** by clinicians, anatomists refer to this opening as the **superior thoracic aperture**. It is bounded anteriorly by the **superior margin of the manubrium** of the sternum, later-ally by the **first rib** and its cartilages, and posteriorly by the **body of the T1** vertebra.

The superior thoracic aperture serves as a route for structures to enter and exit the thorax. Structures that descend from the neck to enter the thorax are the **esophagus, trachea, internal thoracic artery, subclavian and internal jugu-lar veins**, the **vagus**, the **phrenic and cardiac nerves,** and the **sympathetic trunk.** Structures exiting the thorax through the superior aperture are the apex of the **two lungs and the cervical pleura,** the **subclavian artery,** and the **recurrent laryngeal nerves** (see Figure 58-1).

Important anatomical and clinical structures near the superior thoracic aperture are the anterior scalene muscles, which insert on the first rib; and the subclavian vein and artery, which are related anteriorly and posteriorly to insertion of the **ante-rior scalene muscle.** Just superior to the subclavian artery is the lower portion of the **brachial plexus**; the plexus is the major nerve supply to the upper limb. The plexus emerges from between the anterior and middle scalene muscles. Thus both the blood-and-nerve supply to the upper limb lies in or anatomically very close to the superior thoracic aperture.

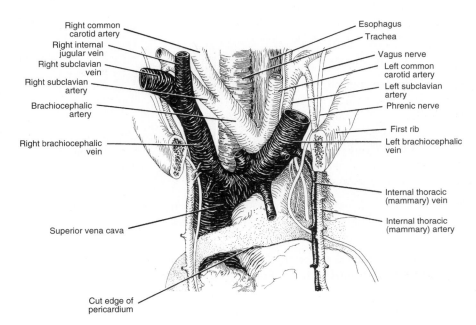

Figure 58-1. Superior mediastinum and root of the neck. [*Reproduced, with permission, from Lindner HH. Clinical Anatomy. East Norwalk, CT: Appleton & Lange, 1989:226 (Figure 17-6)*].

COMPREHENSION QUESTIONS

58.1 Which of the following structures defines the lateral boundary of the superior thoracic aperture?

A. Clavicle

B. Rib 1

C. Rib 2

D. Manubrium

E. Disk between T3 and T4

58.2 You are in anatomy lab and ask a fellow medical student about the course of the right subclavian artery. Your colleague answers that this structure lies immediately posterior to the

A. Clavicle

B. Anterior scalene muscle

C. Middle scalene muscle

D. Posterior scalene muscle

E. Sternocleidomastoid muscle

58.3 You demonstrate the relationship between the brachial plexus and the superior thoracic aperture during the dissection. What portion of the plexus lies closest to the aperture?

A. C5 root

B. C6 root

C. Superior trunk

D. Middle trunk

E. Inferior trunk

ANSWERS

58.1 **B.** Rib 1 forms the lateral boundary of the superior thoracic aperture. The superior margin of the manubrium and the body of T1 form the anterior and posterior boundaries, respectively.

58.2 **B.** The subclavian artery, on both sides, lies immediately posterior to the anterior scalene muscle. The subclavian vein lies immediately anterior to this muscle.

58.3 **E.** The most inferior portion of the brachial plexus listed is the inferior trunk. The other portions of the plexus are more superior.

ANATOMY PEARLS

► The boundaries of the superior thoracic aperture lie from anterior to posterior: superior border of the manubrium, first ribs and cartilages, and the body of T1.

► The subclavian vein and artery lie anterior and posterior, respectively, to the insertion of the anterior scalene muscle as they cross the first rib, and groove it slightly.

► The superior thoracic aperture is tilted inferiorly anterior, allowing the cervical pleura or cupula and lung apex to project superiorly into the neck.

► The most common cause of thoracic outlet syndrome is pressure from a cervical rib.

REFERENCES

Gilroy AM, MacPherson BR, Ross LM. *Atlas of Anatomy,* 2nd ed. New York, NY: Thieme Medical Publishers; 2012:55, 59, 65–66, 74, 89, 99, 593–595, 607–608.

Moore KL, Dalley AF, Agur AMR. *Clinically Oriented Anatomy,* 7th ed. Baltimore, MD: Lippincott Williams & Wilkins; 2014:72–79, 85, 160–168, 721–725, 1012–1017.

Netter FH. *Atlas of Human Anatomy,* 6th ed. Philadelphia, PA: Saunders; 2014: plates 186, 195, 203.

Listing of Cases

Listing by Case Number

Listing by Disorder (Alphabetical)

Listing by Case Number

Listing by Disorder (Alphabetical)

Note: Page numbers followed by a *t* or *f* indicate that the entry is included in a table or figure.

broad ligament, 197
Broca area, 247
bronchial arteries, 101
bronchioles, 100
bronchogenic carcinoma, 76–77
bronchomediastinal trunk, 81
bronchopulmonary segments, 100
bruits, 247
buccinator, 272
buccopharyngeal fascia, 337
Buck fascia, 190
bulbospongiosus muscle, 183, 186, 190
bulbourethral glands, 209, 212
bulimia, 297
bursa
    omental, 148, 153, 363
    of shoulder, 372
    subacromial, 41–42
    subscapular, 39

C
calcium channel blockers, 300
calvaria, 283, 290
Camper fascia, 183
cancer
    breast, 84
    lung, 258
    lymphatics and, 259–260
    metastatic cervical, 196
    testicular, 188
canthus, 342
capillary endothelium, 100–101
caput succedaneum, 282–283
carbamazepine, 270
cardia, 165
cardiac conduction system, 93, 94f
cardiac muscle cells, 96
cardiac nerve, 377
cardiac skeleton, 94
cardiac valves, 94–95
cardiac veins, 114
cardinal ligaments, 198, 216
carotid arteries, 250
    common, 247
    external, 239, 248f, 284, 320
    internal, 247, 284, 301, 304
carotid insufficiency, 246
carotid sheath, 249, 337, 339
carpal bones, 25, 27, 32f
carpal tunnel, 30–32, 35
carpal tunnel syndrome, 30–31
cartilage. *See specific cartilage types*
caruncle, 294
catecholamines, 177
cauda equina, 4, 222

caudal pons, 265
caudate lobes, 158, 161
cavernous sinus, 233, 284, 301
cecal folds and fossae, 143f
cecum, 143
celiac artery, 137, 139, 144, 150, 158, 161, 165, 167
cephalohematoma, 282
cerebellar arteries, 302
cerebellopontine junction, 265
cerebellum, 247
    blood supply to, 304
cerebral aqueduct of Sylvius, 233, 235
cerebral arteries, 290
    anterior, 301–302
    posterior, 302
cerebral cortex, arterial supply to, 302
cerebral decompression, 288
cerebrospinal fluid (CSF), 222–223
    choroid plexus and, 349
    circulation of, 351f
    hydrocephalus and, 348
    meninges and, 232–234, 352
cerebrovascular accident, 246
cerebrum, 247
cervical canal, 197
cervical cytology, 197
cervical dysplasia, 197
cervical lymph nodes, 86
cervical pleura, 377
cervical rib, 377
chest radiograph, 76
    pneumothorax and, 104
chest tube, 105
chicken pox, 226
choanae, 331
choking response, 241
cholecystectomy, 323, 325
cholecystitis, 130
chorda tympani, 264, 266, 268, 309
choroid plexus, 232–233, 349
chronic obstructive pulmonary disease, 105
circle of Willis, 247, 250, 265, 301f, 302, 304
    aneurysm of, 275
    berry aneurysm and, 300
circumcision, 188, 190
circumflex arteries, 65
circumflex branch, 112
circumflex femoral artery, 48
circumflex iliac arteries, 117
cirrhosis, 156
cisterna chili, 78, 177, 259
clavicle, 39, 337
clitoris, 175–176, 183, 186
    penis and, 189